About this Book

"Dr. Zarandi's science training and teaching experience at Caltech has served him—and his readers—well. He has the gift of being able to describe the operating premises of modern scientific thought and their limitations. Whether you are a scientist, scholar or simply someone who takes life seriously, this book will encourage you to reexamine your most basic assumptions about life. Each reader's understanding of the human condition is improved by this insightful collection."
    —**Professor A. Shakouri**, University of California

"Modern science believes—and fears—that by a kind of endless curiosity about the mechanism of material reality, we will eventually encompass—and thus exhaust—the enigma of existence in thought. The articles in this collection make it clear that the corollary is a vain effort to achieve a material utopia whose fulfillment always eludes us. In doing so, they bring a fresh enjoyment to the journey of discovery that science really is."
    —**Professor M. S. Alouini**, University of Minnesota

"Included here are articles by scientists and academicians that explore the limitations of modern science and the experimental method. Is the assumption correct that the technological advances of modern science are incontestable proof of humanity's progress toward the realization of its ultimate well-being? Readers will see not only the practical but also the fundamental limitations of technocracy and the insufficiencies of its image of humanity."
    —**Professor A. K. Ziarani**, Clarkson University

"The essays in this volume are a welcome critique of the hegemonic grasp which modern science holds on contemporary thought, avoiding both simplistic creationist ideology and postmodern relativism. It should be read by all thoughtful scientists and will be a valuable resource for philosophy of science courses."
    —**Dr. William Wroth**

"The essays collected in this volume provide a profound critique of modern scientism but also resist the temptations of postmodernist antirealism and sentimental pietism. While the postmodern critiques of modern science are based on the denial of truth and thus unable to present an intelligible alternative, sentimental pietism and ethicalism criticize the consequences of modern science and technology without grasping their philosophical foundations. The essays go beyond the 'little bit science, little bit ethics' approach, and reassert the urgency of developing a different philosophical framework within which we can make sense of reality, both physical and metaphysical. Highly recommended for those interested in philosophy, science, religion and the environment."
    —**Professor Ibrahim Kalin**, College of the Holy Cross

"Here are essays from respected scholars—both inside and outside the scientific community—who share a clear consensus that physical existence can be understood adequately only as a manifestation of a higher, supraformal reality which is spiritual. Those searching for an alternative to conventional wisdom that can stand up to scrutiny will be well pleased with these essays. Highly recommended."
    —**Professor Fariba Bahrami**, Tehran University

World Wisdom
The Library of Perennial Philosophy

The Library of Perennial Philosophy is dedicated to the exposition of the timeless Truth underlying the diverse religions. This Truth, often referred to as the *Sophia Perennis*—or Perennial Wisdom—finds its expression in the revealed Scriptures as well as the writings of the great sages and the artistic creations of the traditional worlds.

The Perennial Philosophy provides the intellectual principles capable of explaining both the formal contradictions and the transcendent unity of the great religions.

Ranging from the writings of the great sages of the past, to the perennialist authors of our time, each series of our Library has a different focus. As a whole, they express the inner unanimity, transforming radiance, and irreplaceable values of the great spiritual traditions.

Science and the Myth of Progress appears as one of our selections in The Perennial Philosophy series.

The Perennial Philosophy Series

In the beginning of the Twentieth Century, a school of thought arose which has focused on the enunciation and explanation of the Perennial Philosophy. Deeply rooted in the sense of the sacred, the writings of its leading exponents establish an indispensable foundation for understanding the timeless Truth and spiritual practices which live in the heart of all religions. Some of these titles are companion volumes to the Treasures of the World's Religions series, which allows a comparison of the writings of the great sages of the past with the perennialist authors of our time.

Science and
the Myth of Progress

Edited by

Mehrdad M. Zarandi

World Wisdom

Science and the Myth of Progress
© 2003 World Wisdom, Inc.

For complete bibliographic information on the articles
in this anthology, please refer to the page of acknowledgments
placed at the back of the book.

Library of Congress Cataloging-in-Publication Data

Science and the myth of progress / edited by Mehrdad M. Zarandi.
    p. cm. – (The perennial philosophy series)
Includes bibliographical references and index.
ISBN 0-941532-47-X (pbk. : alk. paper)
    1. Science–Philosophy. I. Zarandi, Mehrdad M., 1963- II. Series.
Q175.3.S313 2003
501–dc21

                                                            2003010008

Printed on acid-free paper in Canada

For information address World Wisdom, Inc.
P.O. Box 2682, Bloomington, Indiana 47402-2682

www.worldwisdom.com

The world is a theophany,
an epiphany of things
themselves unseen.

- *A. K. Coomaraswamy*

Contents

Foreword

"The only progress is to doubt of progress"

Nicolas Gomez Davila

This organic collection of essays on science and the myth of progress resembles a mosaic of value, where each part has its function in the whole, while some tesserae play a more pivotal role, giving a precious mark to it. In this case we are mainly speaking of the writings of Guénon, Schuon, Nasr, Burckhardt and some other traditionalist authors, who expose the deep roots of the spiritual crisis of modern man: namely the ills of individualism and rationalism. These two errors spread in an uncontrolled way after the Renaissance period, when the opposition exploded between the "traditional" and "anti-traditional" world views, following a long period of incubation.

For Guénon, the cause of the birth of modern science was "individualism", meaning by this term a *humanistic* reduction of the human being, together with the denial of every supernatural, metaphysical principle. This individualism is the projection at the human level of a more general rationalistic process which has as its result a fragmentation of man's whole view of Reality, which at its unitary height includes—in a normal view—all the multiplicity of universal manifestation. In the course of this subversive process, all supra-sensorial levels of Being—the psychic being finally only an extension of the sensorial—were first relegated to the periphery and then finally excluded, such that only the lower (material) planes of reality are acknowledged in the mind of modern man. Modern science has been rightly called by Nasr "a limited and particular way of knowing things through the observation of their external aspects, of phenomena, and of ratiocination based upon this empirical contact with things". This kind of science and the secular myth of progress are strictly linked, because they reinforce each other with many commonly accepted but false, materialistic beliefs. *Science is seen as the unique, true tool of progress, and progress is the landscape in which science can act with a specific and exact aim.*

As pointed out by the perennialist thinkers in this anthology, within the traditional civilizations there existed different sciences, all perfectly legitimate and true. According to Guénon, traditional sciences are "prolongations or reflections of absolute and principial knowledge". Because they are in "the world of form and multiplicity", there are necessarily several traditional sciences, and they are defined not only by their object, but also by "the standpoint from which the object is envisaged". It was only after man's view of time was transformed from a more qualitative and "cyclic" conception into a mechanical and quantitative flow that people began to imagine a "tendency" in history from the "less" to the "more", and thus our civilization was ready to accept the modern idea of "progress". In this way, "man became the victim of time. . . . The mentality of today seeks to reduce everything to categories connected with time" (Schuon), forgetting that "what is sometimes called the 'tendency of history' is only the law of gravity" (Schuon), which can be symbolized also by the phenomenon of entropy, the progressive increase of disorder in a closed physical system.

The concept of "progress" was also well known in antiquity, but in an altogether different manner than now. By "progress", premodern people meant a relative improvement, limited to specific fields, concerning mainly the material life of man, and without any ideological or pseudo-religious meaning. They spoke about "progresses", with different ends for different people and cultures, in a cyclic becoming of the cosmos where there was also the possibility of "falls" and "regressions". For them, every real improvement—"progress"—was a discontinuous event, a jump in the history of man, often due to either the intervention of the gods or to a gift from them; it was a transient conquest to be defended, and tragedy always lay waiting in ambush. In contrast, after the Middle Ages, slowly a new idea of "Progress"—one completely unknown before—took shape, as a unique, absolute, omnicomprehensive, linear, deterministic and mechanical process, and all this within an increasingly humanistic view of history, culminating in the eighteenth century. Henceforth, apart from a few isolated philosophers, man refused more and more to view the value of earthly life in terms other than those related to his material wellbeing. He lost the consciousness of—and thus also the value of—the "limit" as an ineluctable presence in human life and, more generally, in physical reality. In the modern age, after the rejection of a worldview based on spiritual principles and religious faith, a parodistic inversion of the spiritual world took place, a real metaphysical subversion, confirming that *"Diabolus est deus inversus"*; this was accompanied by a loss of the awareness that every behavior of man is *structurally* "religious", as shown by Mircea Eliade. In the place of the universal Principle, *matter* was transformed into a pseudo-absolute entity and certain aspects of *becoming* came to usurp for modern man the roles of unicity and of infinity, which previously had been recognized as belonging to the supernatural, transcendent dimension: the One and Infinite Absolute. Thus eventually we encounter, instead of several different "progresses" or "sciences", the modern Progress-Science as a unique and totalitarian process tending toward a unique end. Instead of time, space and matter as *indefinite* (quantitative) dimensions, we are presented with the modern pseudo-absolutes of Time, Space, and Matter as supposedly *infinite* (qualitative) realities. They are all the false gods of today, mimicking the true God like apes. In effect, only the existence of the natural level—that of physical and psychic phenomena—is admitted and every other plane of Reality is denied. The "vertical" dimension first becomes neglected—because believed to be untrue and illusory—and then insidiously it is translated into the "horizontal", with all the corresponding distortions, by a totalitarian rationalism, psychologism and biologism.

Modern, or profane, science has consequently deprived nature of its symbolic meaning, thus reducing it to only opaque and heavy matter; whereas in religious cultures, the natural world is viewed as an epiphany of the Sacred, through the analogy and correspondence between the higher and lower levels of Reality. Today "the world is still seen as devoid of a spiritual horizon, not because there is no such horizon present but because the subject who views the contemporary landscape is most often the type of man who lives at the rim of the circle of existence and therefore views all things from the periphery" (Nasr). This man "has projected the externalized and superficial image of himself upon the world" (Nasr). We must not forget that forms—in their most important function, which is symbolic—are a door open toward the supra-formal. The "form", in its higher meaning, is the qualitative (*archetypical*) aspect of nature, its "inner" reality. The Inward—timeless and essential—is the origin of the Outward. The phenomena of nature bear the imprint, the seal of the *supra-rational and immutable*. And as Burckhardt reminds us, the word "cosmos means *order*, implying the ideas of unity and totality".

Modern science has reduced the qualitative aspects of nature to quantitative modalities to better manipulate and dominate matter, following the old program of Francis Bacon. This has led to a situation that resembles more a nightmare than a dream: science is more and more becoming a "techno-science", where even the simple, practical knowledge of daily life is gradually being replaced by a profane lust for power over nature (a "promethean knowledge" according to Schuon). One example typical of the means to this end is the radical oversimplification of every complex reality, with the illusion that, behind the complexity of nature ("complexity" believed to be only seeming), there is a very simple structure. The underlying psychological motivation for this procedure is the vain hope of attaining control of physical reality, the limits of which would have to be acknowledged in the presence of an "irreducible complexity".[1] Such complexity obviously poses a serious hindrance to a goal of control or large scale manipulation because there are so many interferences and interactions—too many to be understood in their global action and effectively managed.

We can repeat, with Guénon, that still today the "*superstition of facts*" is widespread among scientists, but the same facts can always be equally well explained by a variety of different theories, on the basis of various—and in some cases opposite—"preconceived ideas", which are the same thing as the "paradigms" of the epistemologist Thomas Kuhn.[2] For mechanistic science, a living organism, like an animal, plant or human being, is nothing more than a machine composed of material elements or, in other word, a device constructed by the genes to ensure the production of more genes like themselves (among scientists, the socio-biologists express such concepts in the most radical manner). One of the characteristics of modern science is its hyperspecialization, which entails adopting an almost exclusively analytical approach to nature, in comparison to the traditional sciences, in which synthesis was pre-

1 On the concept of "irreducible complexity" in biology see: M. Behe, *Darwin's Black Box*, The Free Press, New York, 1996.
2 T. Kuhn, *The Structure of Scientific Revolutions*, Chicago, University of Chicago Press, 1970; and *The Essential Tension*, Chicago: University of Chicago Press, 1977.

dominant and fundamental. Another aspect is its lack of depth, which is related to its horizontal dispersion. We find a good example of this in the practice of modern medicine. Today the physician is an engineer of human health, a superficial specialist who has invested years in acquiring the professional blinkers which allow him to see only the *part* (the biochemical aspect of the illness) and not the *whole* (the entire man, including body, soul and spirit). This is a sharp contrast to the "medicine man" found in traditional communities who follows a holistic approach to healing. He has a good knowledge not merely of physiology, biology, botany, minerology, chemistry and physics—approached from a point of view in many cases quite different from that of the reductionist sciences—but also of astrology, music, numerology, theology, metaphysics, with a capacity to "understand" in depth and in breadth the meaning of illness.[3] The situation of the modern, spiritually atrophied man—who is like a fading shadow of the "integral" man—is finally hardly surprising.

But within this gloomy picture we also find some rays of light. They constitute nuggets of gold among the grains of sand. We are referring to thinkers and researchers like Denton, Sermonti, Fondi, Chauvin, Thom, Behe, Penrose. Regarding the latter, even if we can agree with some criticisms by William Dembski of his ideas, we consider him a good example of an interesting and stimulating scientist of today, especially for his acceptance of Plato's philosophy: "my sympathies lie strongly with the Platonistic view that mathematical truth is absolute, external, and eternal, and not based on man-made criteria; and that mathematical objects have a timeless existence of their own, not dependent on human society nor on particular physical objects".[4] We would like to end this foreword by quoting a sentence of Ananda K.Coomaraswamy, who, reminding us that *opinion* corresponds to becoming and *truth* to being, writes: "while we are thinking of eternal things, the things that do not change, we are participating in eternity. Eternity is not far away from us, but nearer than time, of which both parts are really far

---

3 M. Lings, *Ancient Beliefs and Modern Superstitions*, Quinta Essentia, Cambridge, 1991, p. 31.

4 R. Penrose, *The Emperor's New Mind*, Penguin Books, New York, 1991, p.116.

away, one far ahead of us and the other far behind us".[5] In practical terms, modern man should free his mind from the *tyranny of matter*, in order to see again the whole of Reality. Only in this way will he be capable of profound change, forgetting all the "modern superstitions" and developing "a lifestyle which accords to material things their proper, legitimate place, which is secondary and not primary" (E. F. Schumacher).

Giovanni Monastra

5 A. K. Coomaraswamy, *Time and Eternity*, Munshiram Manoharlal, New Delhi, 1988, p.129

Preface

There is a domain of reality which is the material world. There is a means of acquiring knowledge about this domain, which is the human mind with its tool of reason. Then there are the many causalities that the human mind seeks to unveil in the material world or physical reality. The modern era has been distinguished by an ever-increasing acquisition of information about the physical universe. This plethora of information and the technological wonders it has produced are inextricably bound up with the idea of human "progress" through material enrichment and empowerment.

What then is "progress"? Can it be "proven"? Any dictionary will tell us that progress is a kind of advancement toward a goal, an improvement, a development from a lower state to a higher one. A simple look into the world of nature will show us many examples of it: in the unfolding of the seasons, in the course of all natural life. But can the idea of progress rightly be applied to the human state as such? Modern thought, beginning roughly with the Renaissance, tells us that it can. Whether implicitly or explicitly, its basic assumptions are the following:

- The centrality of material or physical reality.
- The primacy of the rational faculty and the experimental method as the means of human knowledge.
- The definition of the human vocation, and its ultimate well-being, as the quantifiable understanding of "how" material reality functions.[1]

The above principles have led us to the "age of reason" and to the "age of science"—to a knowledge of the material world by means of human reason applied to the search for the "how" of the world's existence and operation. In this anthology, we want to analyze and

---

1 Galileo stated that, in his search for causalities in the material world, he did not seek for answers to the "why" of phenomena. Instead, he stated that he sought answers to the "how" of phenomena.

reflect on the consequences of this approach. Some of the questions that we seek to answer are:

- Is material or physical reality the only form, or the central form, of reality?
- Is the rational faculty the only faculty, or the central faculty, of human knowledge?
- Is the most profound and complete definition of human nature given by our knowledge of material reality?

The answers to these questions have an importance that is far from being abstract and merely theoretical: for these answers have consequences that reverberate through all aspects of human life and its environment. By our nature, we seek for an inward image of our ultimate goals. Based on our understanding of what the world is and what human beings are, we look for happiness, for a sense of purpose in life and a sense of both wonder and order in the world around us. But if we are basing ourselves on a conception of things that is incomplete, we will not be successful in finding what we seek. We want to know not just the measurable dimension of the "how" of existence; we have a deep thirst for answers in the dimension of its "why." Without answers to the "why" of life, our information on the modalities of its "how" will be of little use to us and may even prove harmful.

There is a great deal currently being written about the limitations entailed in the three principles outlined above. Many are involved in the task of examining the practical consequences of technocracy for our lives and our environment. Through the essays collected here, we wish to examine the modern image of humanity and the universe from a more philosophical point of view. Presenting the problem is the first step. There are essays here from several philosophers, scientists, an economist and an agriculturist-poet. They are all asking the same question: Are there limits to what the science of the material world can offer us, and thus to "progress"? And their unanimous answer is "yes."

Seeing a problem, we normally seek for a solution. Where should we look? The purpose of these chapters is less to insist upon one particular "answer" than to offer alternative points of departure to the premises of modernism that permeate our culture so thoroughly. And again, there is a consensus among our authors. Human beings are capable of discerning between knowledge and ignorance; and within the domain of knowledge between that which is essential and that which is non-essential. Both have their importance, but if words have any meaning, we have to give primacy to what is essential. Said in another way, there is Truth and then there are truths. There are many truths that pertain to the "how" of existence, and modern science offers them in abundance. But these should not be confused with Truth, which pertains to the "why" of existence and to its "final cause" as described by Aristotle—and even more profoundly by Plato. This Truth is the domain of spiritual reality and the traditional wisdoms which have been its repositories from time immemorial. We may ignore spiritual Truth, or even temporarily suppress it; but we do so at our own peril, for It is finally as ineluctable as the forces of nature that manifest It.

The word "science" comes from the Latin *scientia* or *sciens*, which means, "having knowledge." Science bases itself on objectivity: one should weigh all the data before arriving at a conclusion. Modern science has incontestably brought to light a dazzling array of information unknown to previous generations. But the contemporary belief in an endless progress tends toward an almost total rejection of spiritual wisdom's worldviews as being naïve, outmoded and contrary to empirical evidence. Is it not possible that the glaring spectacle of technology's prodigies has caused us to lose sight of something which is in fact crucial? The purpose of this anthology is to provide access to information that is worthy of our consideration in evaluating this question.

Mehrdad M. Zarandi
Bloomington, Indiana
October 2002

1

In the Wake of the Fall

Frithjof Schuon

In antiquity and in the Middle Ages man was "objective" in the sense that his attitude was still largely determined by "objects,"[1] by objects on the plane of ideas as well as by objects of the senses; he was very far from the relativism of modern man who impairs objective reality by reducing it to accidents of nature lacking in significance and in symbolic quality; and he was equally far from a "psychologism" which calls into question the value of the subject, the knower, and thus in effect destroys the very idea of intelligence. To speak of "objects on the plane of ideas" is not a contradiction, since a concept, while it is evidently a subjective phenomenon insofar as it is a mental phenomenon, is at the same time, like every sensory phenomenon, an object in relation to the subject who takes cognizance of it; truth comes in a sense from outside, it is offered to the subject who may accept it or not accept it. Held fast as it were to the objects of his knowledge or of his faith, ancient man was little disposed to grant a determining role to psychological contingencies; his inner reactions, whatever their intensity, were related to an object and thereby had in his consciousness a more or less objective cast. The object as such—the object envisaged in all its objectivity—was the real, the basic, the immutable thing, and in grasping the object, he had hold of the subject; the subject was guaranteed by the object. This is, of course, always the case with many men, and in certain respects even with every sane man; but the aim here—at the risk of seeming to propound truisms—is to indicate, despite the fact that it can only be done approximately, the outlines of points of

---

1 In current usage, the words "objective" and "objectivity" often carry the meaning of impartiality, but it goes without saying that in the present context they are not used in that derivative and secondary sense.

1

view that are in the nature of things complex. In any case, to be too easily satisfied by attention to the subject is to betray the object; the men of old would have had the impression of denaturing or losing the object if they had paid too much attention to the subjective pole of consciousness. It was only from the time of the Renaissance that the European became "reflexive," and so in a certain way subjective; it is true that such a reflexivity can in its turn have a perfectly objective quality, just as an idea received from without can have a subjective character owing to some bias of interest or feeling in the subject, but this aspect of the matter is not relevant here; what is relevant is that at the Renaissance man began to analyze mental reflections and psychic reactions and thus to be interested in the "subject" pole to the detriment of the "object" pole; in becoming "subjective" in this sense, he ceased to be symbolist and became rationalist, since reason is the thinking ego. It is this that explains the psychological and descriptive tendencies of the great Spanish mystics, tendencies which have been wrongly taken as evidence of a superiority and as a kind of norm.

This transition from objectivism to subjectivism reflects and repeats in its own way the fall of Adam and the loss of Paradise; in losing a symbolist and contemplative perspective, founded both on impersonal intelligence and on the metaphysical transparency of things, man has gained the fallacious riches of the ego; the world of divine images has become a world of words. In all cases of this kind, heaven—or a heaven—is shut off from above us without our noticing the fact and we discover in compensation an earth long unappreciated, or so it seems to us, a homeland which opens its arms to welcome its children and wants to make us forget all Lost Paradises; it is the embrace of *Mâyâ*, the sirens' song; *Mâyâ*, instead of guiding us, imprisons us. The Renaissance thought that it had discovered man, whose pathetic convulsions it admired; from the point of view of laicism in all its forms, man as such had become to all intents and purposes good, and the earth too had become good and looked immensely rich and unexplored; instead of living only "by halves" one could at last live fully, be fully man and fully on earth; one was no longer a kind of half-angel, fallen and exiled; one had become a whole being, but by the downward path. The Reformation, whatever certain of its tendencies may have been, had as an overall result the relegation of God to Heaven—to a Heaven henceforth distant and more and more neutralized—on the pretext that God keeps close to us "through Christ" in a sort of Biblical atmosphere, and that He resembles us as we resemble Him. All this brought with it an apparently miraculous enrichment of the aspect of things as "subject" and "earth," but a prodigious impoverishment in their aspect as "object" and "Heaven." At the time of the Revolution of the late eighteenth century, the earth had become definitely and exclusively the goal of man; the "Supreme Being" was merely a "consolation" and as such a target for ridicule; the seemingly infinite multitude of things on earth called for an infinity of activities, which furnished a pretext for rejecting contemplation and with it repose in "being" and in the profound nature of things; man was at last free to busy himself, on the hither side of all transcendence, with the discovery of the terrestrial world and the exploitation of its riches; he was at last rid of symbols, rid of metaphysical transparency; there was no longer anything but the agreeable or the disagreeable, the useful or the useless, whence the anarchic and irresponsible development of the experimental sciences. The flowering of a dazzling "culture" which took place in or immediately after these epochs, thanks to the appearance of many men of genius, seems clearly to confirm the impression, deceptive though it be, of a liberation and a progress, indeed of a "great period"; whereas in reality this development represents no more than a compensation on a lower plane such as cannot fail to occur when a higher plane is abandoned.

Once Heaven was closed and man was in effect installed in God's place, the objective measurements of things were, virtually or actually, lost. They were replaced by subjective measurements, purely human and conjectural pseudo-values, and thus man became involved in a movement of a kind that cannot be halted, since, in the absence of celestial and stable values, there is no longer any reason for calling a halt, so that in the end a stage is reached at which human values are replaced by infra-human values, up to a point at which the very idea of truth is abolished. The mitigating circumstances in such cases—for they are always present, at any rate for some individuals—consist in the fact that, on the verge of every new fall, the order then existing shows a maximum of abuse and corruption, so that the temptation to prefer an apparently clean error to an outwardly soiled truth is particularly strong. In a traditional civilization, the mundane element does all it can to compromise in the eyes of the majority the principles governing that civilization; the majority itself is only too prone to be worldly, its worldliness is not however aristocratic and light-hearted, but ponderous and pedantic. It is not the people who are the victims of theocracy, it is on the contrary theocracy that is the victim, first of aristocratic worldlings and finally of the masses, who begin by being seduced and end in revolt. The European monarchs of the nineteenth century made almost desperate efforts to dam the mounting tide of a democracy of which they had already, partially and despite themselves, become representatives. But these efforts were doomed to be vain in default of the one counterweight that could have re-established stability, and that could only be religion, sole source of the legitimacy and the power of princes. They fought to maintain an order in principle religious, but the forms in which this order was presented disavowed it themselves; the very apparel of kings, and all the other forms among which they lived, proclaimed doubt, a spiritual "neutralism," a dimming of faith, a bourgeois and down-to-earth worldliness. This was already true, to a lesser degree, in the eighteenth century, in which the arts of dress, architecture and craftsmanship expressed, if not exactly democratic tendencies, at least a worldliness lacking in greatness and strangely insipid. In this incredible age all men looked like lackeys—the nobles all the more so for being nobles—and a rain of rice-powder seemed to have fallen on to a world of dreams; in this half-gracious and half-despicable universe of marionettes, the Revolution merely took advantage of the previous suicide of the religious outlook and of greatness, and could not but break out; the world of wigs was much too unreal. Analogous remarks, suitably attenuated to conform to eminently different conditions, apply to the Renaissance and even to the end of the Middle Ages; the causes of the descent are always the same when seen in relation to absolute values. What is sometimes called the "tendency of history" is only the law of gravity.

To state that the values of ancient man were celestial and static, amounts to saying that man then still lived "in space"; time was merely the contingency that corroded all things; in the face of that contingency values that are so to speak "spatial," that is, permanent by virtue of their intemporal finality, had always to assert themselves anew. Space symbolizes origin and immutability; time is the decadence which carries us away from the origin while at the same time leading us towards the Messiah, the great Liberator, and towards the meeting with God. In rejecting or losing celestial values, man became the victim of time; in inventing machines which devour duration man has torn himself away from the peacefulness of space and thrown himself into a whirlpool from which there is no escape.

The mentality of today seeks in fact to reduce everything to categories connected with time; a work of art, a thought, a truth have no value in themselves and independently of any historical classification, but their value is always related to the time in which they are rightly or wrongly placed; everything is considered as the expression of a "period" and not as having a timeless and intrinsic value; and this is entirely in conformity with modern relativism, and with a psychologism or biologism that destroys essential values. In order to "situate" the doctrine of a scholastic, for example, or even of a Prophet, a "psychoanalysis" is prepared—it is needless to emphasize the monstrous impudence implicit in such an attitude—and with wholly mechanical and perfectly unreal logic the "influences" to which this doctrine has been subject are laid bare. There is no hesitation in attributing to saints, in the course of this process, all kinds of artificial and even fraudulent, conduct; but it is obviously forgotten, with satanic inconsequence, to apply the same principle to oneself, and to explain one's own supposedly "objective" position by psychoanalytical considerations; sages are treated as being sick men and one takes oneself for a god. In the same range of ideas, it is shamelessly asserted that there are no primary ideas; that they are due only to prejudices of a grammatical order—and thus to the stupidity of the sages who were duped by them—and that their only effect has been to sterilize "thought" for thousands of years, and so on and so forth; it is a case of expressing a maximum of absurdity with a maximum of subtlety. For procuring a pleasurable sensation of important accomplishment there is nothing like the conviction of having invented gunpowder or of having stood Christopher Columbus' egg on its point. This philosophy derives all it has in the way of originality from what, in effect, is nothing but a hatred of God; but since it is impossible to abuse directly a God in whom one does not believe, one abuses Him indirectly through the laws of nature,[2] and one goes so far as to disparage the very form of man and his intelligence, the very intelligence one thinks with and

2 A contemporary writer whose name does not come to mind has written that death is something "rather stupid," but this small impertinence is in any case a

abuses with. There is however no escape from the immanent Truth: "The more he blasphemes," says Meister Eckhart, "the more he praises God."

Mention has already been made of the passage from objectivity to reflexive subjectivity—a phenomenon pointed out by Maritain— and at the same time the ambiguous character of this development has been emphasized. The fatal result of a "reflexivity" that has become hypertrophied is an exaggerated attention to verbal subtleties which makes a man less and less sensitive to the objective value of formulations of ideas; a habit has grown up of "classifying" everything without rhyme or reason in a long series of superficial and often imaginary categories, so that the most decisive—and intrinsically the most evident—truths are unrecognized because they are conventionally relegated into the category of things "seen and done with," while ignoring the fact that "to see" is not necessarily synonymous with "to understand"; a name like that of Jacob Boehme, for example, means theosophy, so "let us move on." Such propensities hide the distinction between the "lived vision" of the sage and the mental virtuosity of the profane "thinker"; everywhere we see "literature," nothing but "literature," and what is more, literature of such and such a "period." But truth is not and cannot be a personal affair; trees flourish and the sun rises without anyone asking who has drawn them forth from the silence and the darkness, and the birds sing without being given names.

In the Middle Ages there were still only two or three types of greatness: the saint and the hero, and also the sage; and then on a lesser scale and as it were by reflection, the pontiff and the prince; as for the "genius" and the "artist," those glories of the lay universe, their like was not yet born. Saints and heroes are like the appearance of stars on earth; they reascend after their death to the firmament, to their eternal home; they are almost pure symbols, spiritual signs only provisionally detached from the celestial iconostasis in which they have been enshrined since the creation of the world.

---

characteristic example of the mentality in question. The same outlook—or the same taste—gave rise to a remark, met with a little time ago, that a certain person had perished in an "idiotic accident." It is always nature, fate, the will of God, objective reality, which is pilloried; it is subjectivity that sets itself up as the measure of things, and what a subjectivity!

*

Modern science, as it plunges dizzily downwards, its speed increasing in geometrical progression towards an abyss into which it hurtles like a vehicle without brakes, is another example of that loss of the "spatial" equilibrium characteristic of contemplative and still stable civilizations. This criticism of modern science—and it is by no means the first ever to be made—is made not on the grounds that it studies some fragmentary field within the limits of its competence, but on the grounds that it claims to be in a position to attain to total knowledge, and that it ventures conclusions in fields accessible only to a supra-sensible and truly intellective wisdom, the existence of which it refuses on principle to admit. In other words, the foundations of modern science are false because, from the "subject" point of view, it replaces Intellect and Revelation by reason and experiment, as if it were not contradictory to lay claim to totality on an empirical basis; and its foundations are false too because, from the "object" point of view, it replaces the universal Substance by matter alone, either denying the universal Principle or reducing it to matter or to some kind of pseudo-absolute from which all transcendence has been eliminated.

In all epochs and in all countries there have been revelations, religions, wisdoms; tradition is a part of mankind, just as man is a part of tradition. Revelation is in one sense the infallible intellection of the total collectivity, insofar as this collectivity has providentially become the receptacle of a manifestation of the universal Intellect. The source of this intellection is not of course the collectivity as such, but the universal or divine Intellect insofar as it adapts itself to the conditions prevailing in a particular intellectual or moral collectivity, whether it be a case of an ethnic group or of one determined by more or less distinctive mental conditions. To say that Revelation is "supernatural" does not mean that it is contrary to nature insofar as nature can be taken to represent, by extension, all that is possible on any given level of reality, it means that Revelation does not originate at the level to which, rightly or wrongly, the epithet "natural" is normally applied. This "natural" level is precisely that of physical causes, and hence of sensory and psychic phenomena considered in relation to those causes.

If there are no grounds for finding fault with modern science insofar as it studies a realm within the limits of its competence—the

7

precision and effectiveness of its results leave no room for doubt on this point—one must add this important reservation, namely, that the principle, the range and the development of a science or an art is never independent of Revelation or of the demands of spiritual life, not forgetting those of social equilibrium; it is absurd to claim unlimited rights for something in itself contingent, such as science or art. By refusing to admit any possibility of serious knowledge outside its own domain, modern science, as has already been said, claims exclusive and total knowledge, while making itself out to be empirical and non-dogmatic, and this, it must be insisted, involves a flagrant contradiction; a rejection of all "dogmatism" and of everything that must be accepted a priori or not at all is simply a failure to make use of the whole of one's intelligence.

Science is supposed to inform us not only about what is in space but also about what is in time. As for the first-named category of knowledge, no one denies that Western science has accumulated an enormous quantity of observations, but as for the second category, which ought to reveal to us what the abysses of duration hold, science is more ignorant than any Siberian shaman, who can at least relate his ideas to a mythology, and thus to an adequate symbolism. There is of course a gap between the physical knowledge—necessarily restricted—of a primitive hunter and that of a modern physicist; but measured against the extent of knowable things, that gap is a mere millimeter.

Nevertheless, the very precision of modern science, or of certain of its branches, has become seriously threatened, and from a wholly unforeseen direction, by the intrusion of psychoanalysis, not to mention that of "surrealism" and other systematizations of the irrational; or again by the intrusion of existentialism, which indeed belongs strictly speaking not so much to the domain of the irrational as to that of the unintelligent.[3] A rationality that claims self-sufficiency cannot fail to provoke such interferences, at any rate at its vulnerable points such as psychology or the psychological—or "psychologizing"—interpretation of phenomena which are by definition beyond its reach.

---

3 That is to say if one applies the intellectual norms properly applicable in this case, since it is a question of "philosophy."

It is not surprising that a science arising out of the fall—or one of the falls—and out of an illusory rediscovery of the sensory world should also be a science of nothing but the sensory, or what is virtually sensory,[4] and that it should deny everything which surpasses that domain, thereby denying God, the next world and the soul,[5] and this presupposes a denial of the pure Intellect, which alone is capable of knowing everything that modern science rejects. For the same reasons it also denies Revelation, which alone rebuilds the bridge broken by the fall. According to the observations of experimental science, the blue sky which stretches above us is not a world of bliss, but an optical illusion due to the refraction of light by the atmosphere, and from this point of view, it is obviously right to maintain that the home of the blessed does not lie up there. Nevertheless it would be a great mistake to assert that the association of ideas between the visible heaven and celestial Paradise does not arise from the nature of things, but rather from ignorance and ingenuousness mixed with imagination and sentimentality; for the blue sky is a direct and therefore adequate symbol of the higher— and supra-sensory—degrees of Existence; it is indeed a distant reverberation of those degrees, and it is necessarily so since it is truly a symbol, consecrated by the sacred Scriptures and by the unanimous intuition of peoples.[6] A symbol is intrinsically so concrete and so efficacious that celestial manifestations, when they occur in our sensory world, "descend" to earth and "reascend" to Heaven; a symbolism accessible to the senses takes on the function of the supra-sensible reality which it reflects. Light-years and the relativity of the space-time relationship have absolutely nothing to do with the perfectly "exact" and "positive" symbolism of appearances and its connection at once analogical and ontological with the celestial or angelic orders. The fact that the symbol itself may be no more than an optical illusion in no way impairs its precision or its efficacy, for all appearances, including those of space and of the galaxies, are strictly speaking only illusions created by relativity.

---

4 This distinction is necessary to meet the objection that science operates with elements inaccessible to our senses.

5 Not that all scientists deny these realities, but science denies them, and that is quite a different thing.

6 The word "symbol" implies "participation" or "aspect," whatever difference of level may be involved.

One of the effects of modern science has been to give religion a mortal wound, by posing in concrete terms problems which only esoterism can resolve; but these problems remain unresolved, because esoterism is not listened to, and is listened to less now than ever. Faced by these new problems, religion is disarmed, and it borrows clumsily and gropingly the arguments of the enemy; it is thus compelled to falsify by imperceptible degrees its own perspective, and more and more to disavow itself. Its doctrine, it is true, is not affected, but the false opinions borrowed from its repudiators corrode it cunningly "from within"; witness, for example, modernist exegesis, the demagogic leveling down of the liturgy, the Darwinism of Teilhard de Chardin, the "worker-priests," and a "sacred art" obedient to surrealist and "abstract" influences. Scientific discoveries prove nothing to contradict the traditional positions of religion, of course, but there is no one at hand to point this out; too many "believers" consider, on the contrary, that it is time that religion "shook off the dust of the centuries," which amounts to saying, that it should "liberate" itself from its very essence and from everything which manifests that essence. The absence of metaphysical or esoteric knowledge on the one hand, and the suggestive force emanating from scientific discoveries as well as from collective psychoses on the other, make religion an almost defenseless victim, a victim that even refuses more often than not to make use of the arguments at its disposal. It would nevertheless be easy, instead of slipping into the errors of others, to demonstrate that a world fabricated by scientific influences tends everywhere to turn ends into means and means into ends, and that it results either in a mystique of envy, bitterness and hatred, or in a complacent shallow materialism destructive of qualitative distinctions. It could be demonstrated too that science, although in itself neutral—for facts are facts—is nonetheless a seed of corruption and annihilation in the hands of man, who in general has not enough knowledge of the underlying nature of Existence to be able to integrate—and thereby to neutralize—the facts of science in a total view of the world; that the philosophical consequences of science imply fundamental contradictions; and that man has never been so ill-known and so misinterpreted as from the moment when he was subjected to the "x-rays" of a psychology founded on postulates that are radically false and contrary to his nature.

Modern science represents itself in the world as the principal, or as the only purveyor of truth; according to this style of certainty to know Charlemagne is to know his brain-weight and how tall he was. From the point of view of total truth—let it be said once more—it is a thousand times better to believe that God created this world in six days and that the world beyond lies beneath the flat surface of the earth or in the spinning heavens, than it is to know the distance from one nebula to another without knowing that phenomena merely serve to manifest a transcendent Reality which determines us in every respect and gives to our human condition its whole meaning and its whole content. The great traditions moreover, aware that a promethean knowledge must lead to the loss of the essential and saving truth, have never prescribed or encouraged any such accumulating of wholly external items of knowledge, for it is in fact mortal to man. It is currently asserted that such and such a scientific achievement "does honor to the human race," together with other futilities of the same kind, as if man could do honor to his nature otherwise than by surpassing himself, and as if he could surpass himself except in a consciousness of the absolute and in sanctity.

In the opinion of most men today, experimental science is justified by its results, which are in fact dazzling from a certain fragmentary point of view, but one readily loses sight not only of the decided predominance of bad results over good, but also of the spiritual devastation inherent in the scientific outlook, a priori and by its very nature, a devastation for which its positive results—always external and partial—can never compensate. In any event, it savors of temerity in these days to dare to recall the most forgotten of Christ's sayings: "For what shall it profit a man, if he shall gain the whole world, and lose his own soul?" (Mark 8:36).

*

If the unbeliever recoils from the idea that all his actions will be weighed, that he will be judged and perhaps condemned by a God whom he cannot grasp, that he will have to expiate his faults or even simply his sin of indifference, it is because he has no sense of immanent equilibrium, and no sense of the majesty of Existence, and of the human state in particular. To exist is no small matter; the proof is that no man can extract from nothingness a single speck of dust;

similarly, consciousness is not nothing; we cannot bestow the least spark of it on an inanimate object. The hiatus between nothingness and the least of objects is absolute, and in the last analysis this absoluteness is that of God.[7]

What is outrageous in those who assert that "God is dead" or even "buried"[8] is that in doing so they inevitably put themselves in the place of that which they deny: whether they want to or not, they fill the vacuum left by the loss of the notion of God with psychological constructions, and this confers on them provisionally—and paradoxically—a false superiority and even a kind of pseudo-absoluteness, or a kind of false realism stamped with icy loftiness or if need be with false modesty. Thenceforth their existence—and that of the world—is terribly lonely when faced with the vacuum created by the "inexistence of God";[9] it is the world and it is themselves—they who are the brains of the world—who henceforth carry the whole weight of universal Being instead of having the possibility of resting in it, as is demanded by human nature and above all by truth. Their poor individual existence—as distinct from Existence as such insofar as they participate in it, which moreover appears to them "absurd," if they have any idea of it at all[10]—their existence is condemned to a kind of divinity, or rather to a phantom of divinity, whence the appearance of superiority already mentioned, a posed and polished ease too often combined with a charity steeped in bitterness and in reality set against God.

The artificial isolation in question accounts moreover for the cult of "nothingness" and of "anguish," as well as for the astonishing notion of liberation by action, and even by "dedication" to action.

---

7 It should not be forgotten that God as Beyond-Being, or supra-personal Self, is absolute in an intrinsic sense, while Being or the divine Person is absolute extrinsically, that is, in relation to His manifestation or to creatures, but not in Himself, nor with respect to the Intellect which "penetrates the depths of God."

8 There are Catholics who do not hesitate to hold such views about the Greek Fathers and the Scholastics, doubtless in order to compensate for a certain "inferiority complex."

9 In reality God is indeed not "existent" in the sense that He cannot be brought down to the level of the existence of things. In order to make it clear that this reservation implies no kind of privation it would be better to say that God is "non-inexistent."

10 In any case the idea is restricted to the field of perception of the world and of things, and is therefore quite indirect.

When man is deprived of the divine "existentiation" or when he believes himself so to be, he must find something to take its place, on pain of collapsing into his own nothingness, and he does so by substituting for "existence" precisely this kind of "dedication" to action.[11] In other words, his imagination and his feelings capitulate to the ideal of the machine; for the machine has no value except by virtue of what it produces, and so man exists only by virtue of what he does, and not of what he is; but a man defined by action is no longer man, he is a beaver or an ant.

In the same line of thought, attention must be drawn to the current search for false absolutes on all planes, whence the silly theatricality of modern artists; ancient man, who had a sense of the relativity of values and who put everything in its place, appears to be mediocre by comparison, easily satisfied and hypocritical. The mystical fervor that is a part of human nature is deflected from its normal objects and squandered on absurdities; it is put into a still life or a play, when it is not applied to the trivialities which characterize the reign of the machine and of the masses.

Independently of doctrinal atheism and of cultural peculiarities, modern man moves in the world as if existence were nothing, or as if he had invented it; in his eyes it is a commonplace thing like the dust beneath his feet—more especially as he has no consciousness of the Principle at once transcendent and immanent—and he makes use of it with assurance and inadvertence in a life that has been de-consecrated into meaninglessness. Everything is conceived through the haze of a tissue of contingencies, relationships, prejudices; no phenomenon is any longer considered in itself, in its being, and grasped at its root; the contingent has usurped the rank of the absolute; man scarcely reasons any more except in terms of his imagination falsified by ideologies on the one hand and by his artificial surroundings on the other. But the eschatological doctrines, however exaggerated they may appear to the sensibilities of those whose only Gospel is their own materialism and dissipation

---

11 It is forgotten that the sages or philosophers who have determined the intellectual life of mankind for hundreds or thousands of years—the Prophets not being now under consideration—were in no way "dedicated to action," or rather that their "dedication" was in their work, which is fully sufficient; to think otherwise is to seek to reduce intelligence or contemplation to action, and that comes well into line with existentialism.

and whose life is nothing but a flight before God, are in fact a true yardstick for man's cosmic situation; what the Revelations ask of us and what Heaven imposes or inflicts on us is what we are in reality, whether we think so or not; we know it in our heart of hearts, if only we can detach ourselves a little from the monstrous accumulation of false images entrenched in our minds. What we need is to become once again capable of grasping the value of existence and, amid the multitude of phenomena, the meaning of man; we must once again find the measure of the real! The degree of our understanding of man can be measured by our reactions to what religions teach, or to what our own religion teaches, about the hereafter.

There is something in man which can conceive the Absolute and even attain it and which, in consequence, is absolute. On this basis one can assess the extent of the aberration of those to whom it seems perfectly natural to have the right or the chance to be man, but who wish to be man without participating in the integral nature of man and the attitudes it implies. Needless to say, the possibility of denying itself, paradoxical though it appears, is also a part of human nature—for to be man is to be free in a "relatively absolute" sense—much in the same way as it is humanly possible to accept error or to throw oneself into an abyss.

It has already been pointed out that "unbelievers" no longer have the sense either of nothingness or of existence, that they no longer know the value of existence, and never look at it in relation to the nothingness from which it is miraculously detached. Miracles in the usual sense of the word are in effect only particular variants of this initial miracle—everywhere present—the miracle of the fact of existence; the miraculous and the divine are everywhere; it is the truly human outlook that is absent.

When all is said and done there are only three miracles: existence, life, intelligence; with intelligence, the curve springing from God closes on itself, like a ring which in reality has never been parted from the Infinite.

*

When the modern world is contrasted with traditional civilizations, it is not simply a question of seeking the good things and the bad things on one side or the other; good and evil are everywhere, so that it is essentially a question of knowing on which side the more important good, and on which side the lesser evil, is to be found. If someone says that such and such a good exists outside tradition, the answer is: no doubt, but one must choose the most important good, and it is necessarily represented by tradition; and if someone says that in tradition there exists such and such an evil, the answer is: no doubt, but one must choose the lesser evil, and again it is tradition that embodies it. It is illogical to prefer an evil which involves some benefits to a good which involves some evils.

Nevertheless, to confine oneself to admiring the traditional worlds is still to stop short at a fragmentary point of view, for every civilization is a "two-edged sword"; it is a total good only by virtue of those invisible elements that determine it positively. In certain respects, every human society is bad; if its transcendent character is wholly eliminated—which amounts to dehumanizing it since an element of transcendence is essential to man though always dependent on his free consent—then the whole justification of society's existence is removed at the same time, and there remains only an ant-heap, in no way superior to any other ant-heap, since the needs of life and in consequence the right to life remain everywhere the same, whether the life be that of men or of insects. It is one of the most pernicious of errors to believe, firstly, that the human collectivity as such represents an unconditional or absolute value, and secondly that the well-being of this collectivity represents any such value or any such end in itself.

Religious civilizations, regarded as social phenomena and independently of their intrinsic value—though there is no sharp dividing line between the two—are, despite their inevitable imperfections, like sea-walls built to stem the rising tide of worldliness, of error, of subversion, of the fall and its perpetual renewal. The fall is more and more invasive, but it will be conquered in its turn by the final irruption of the divine fire, that very fire of which the religions are and always have been the earthly crystallizations. The rejection of the traditional religious frameworks on account of human abuses amounts to an assertion that the founders of religion did not know what they were doing, as well as that abuses are not inherent in human nature, and that they are therefore avoidable even in societies counting millions of men, and that they are avoidable through purely human means; no more flagrant contradiction than this could well be imagined.

*

In a certain sense, Adam's sin was a sin arising from in-quisitiveness, if such an expression be admissible. Originally, Adam saw contingencies in the aspect of their relationship to God and not as independent entities. Anything that is considered in that rela-tionship is beyond the reach of evil; but the desire to see contin-gency as it is in itself is a desire to see evil; it is also a desire to see good as something contrary to evil. As a result of this sin of inquisi-tiveness—Adam wanted to see the "other side" of contingency—Adam himself and the whole world fell into contingency as such; the link with the divine Source was broken and became invisible; the world became suddenly external to Adam, things became opaque and heavy, they became like unintelligible and hostile frag-ments. This drama is always repeating itself anew, in collective history as well as in the life of individuals.

A meaningless knowledge, a knowledge to which we have no right either by virtue of its nature, or of our capacities, and there-fore by virtue of our vocation, is not a knowledge that enriches, but one that impoverishes. Adam had become poor after having acquired knowledge of contingency as such, or of contingency insofar as it limits.[12] We must distrust the fascination which an abyss can exert over us; it is in the nature of cosmic blind-alleys to seduce and to play the vampire; the current of forms does not want us to escape from its hold. Forms can be snares just as they can be symbols and keys; beauty can chain us to forms, just as it can also be a door opening towards the formless.

Or again, from a slightly different point of view: the sin of Adam consists in effect of having wished to superimpose something on existence, and existence was beatitude; Adam thereby lost this beat-itude and was engulfed in the anxious and deceptive turmoil of superfluous things.[13] Instead of reposing in the immutable purity of

---

12 A *hadîth* says: "I seek refuge with God in the face of a science which is of no use to me," and another: "One of the claims to nobility of a Muslim rests on not paying attention to what is not his concern." Man must remain in primordial inno-cence, and not seek to know the universe in detail. This thirst for knowledge—as the Buddha said—holds man to the *samsâra*.

13 Compare: "You are dominated by the desire to possess more and more." (Koran 102:1)

Existence, fallen man is drawn into the dance of things that exist, and they, being accidents, are delusive and perishable. In the Christian cosmos, the Blessed Virgin is the incarnation of this snow-like purity; She is inviolable and merciful like Existence or Substance; God in assuming flesh brought with Him Existence, which is as it were His Throne; He caused it to precede Him and He came into the world by its means. God can enter the world only through virgin Existence.

*

The problem of the fall evokes the problem of the universal theophany, the problem that the world presents. The fall is only one particular link in this process; moreover it is not everywhere presented as a "shortcoming" but in certain myths it takes the form of an event unconnected with human or angelic responsibility. If there is a cosmos, a universal manifestation, there must also be a fall or falls, for to say "manifestation" is to say "other than God" and "separation."

On earth, the divine Sun is veiled; as a result the measures of things become relative, and man can take himself for what he is not, and things can appear to be what they are not; but once the veil is torn, at the time of that birth which we call death, the divine Sun appears; measures become absolute; beings and things become what they are and follow the ways of their true nature.

This does not mean that the divine measures do not reach this world, but they are as it were "filtered" by its existential shell; previously they were absolute but they become relative, hence the floating and indeterminate character of things on earth. The star which is our sun is none other than Being seen through this carapace; in our microcosm the Sun is represented by the heart.[14] It is because we live in all respects in such a carapace that we have need—that we may know who we are and whither we are going—of that cosmic cleavage which constitutes Revelation; and it could be

---

14 And the moon is the brain, which is identified macrocosmically—if the sun is Being—with the central reflection of the Principle in manifestation, a reflection susceptible to "waxing and waning" in accordance with its contingent nature and therefore also with cyclic contingencies. These correspondences are of great complexity—a single element can take on various significations—they can therefore only be mentioned in passing. It is sufficient

pointed out in this connection that the Absolute never consents to become relative in a total and uninterrupted manner.

In the fall, and in its repercussions through duration, we see the element of "absoluteness" finally devoured by the element of "contingency"; it is in the nature of the sun to be devoured by the night, just as it is in the nature of light to "shine in the darkness" and not to be "comprehended." Numerous myths express this cosmic fatality, inscribed in the very nature of what could be called the "reign of the demiurge."

The prototype of the fall is in fact the process of universal manifestation. The ideas of manifestation, projection, "alienation," egress, imply those of regression, reintegration, return, apocatastasis; the error of the materialists—whatever subtleties they may employ in seeking to dissolve the conventional and now "obsolete" idea of matter—is to take matter as their starting point as if it were a primordial and stable fact, whereas it is only a movement, a sort of transitory contraction of a substance that is in itself inaccessible to our senses. The matter we know, with all that it comprises, is derived from a supra-sensory and eminently plastic proto-matter; it is in this protomatter that the primordial terrestrial being is reflected and "incarnated"; in Hinduism this truth is affirmed in the myth of the sacrifice of *Purusha*. Because of the tendency to segmentation inherent in this proto-matter, the divine image was broken and diversified; but creatures were still, not individuals who tear one another to pieces, but contemplative states derived from angelic models and, through those models, from divine Names. It is in this sense that it could be said that in Paradise sheep lived side by side with lions; but in such a case only "hermaphrodite" prototypes—of supra-sensory spherical form—are in question, divine possibilities issuing from the qualities of "clemency" and of "rigor," of "beauty" and of "strength," of "wisdom" and of "joy." In this proto-material *hyle* occurred the creation of species and of man, a creation resembling the "sudden crystallization of a supersaturated chemical solu-

---

to add that the sun itself also of necessity represents the divine Spirit manifested, and that it is on this account that it must "wane" in setting and "wax" in rising; it gives light and heat because it is the Principle, and it sets because it is but the manifestation of the Principle; the moon from this point of view is the peripheral reflection of that manifestation. Christ is the sun, and the Church is the moon; "it is expedient for you that I go away" (John 16:7) but the "Son of man will come again."

tion."[15] After the "creation of Eve"—the bipolarization of the primordial "androgyne"—there occurred the "fall," namely the "exteriorization" of the human couple, which brought in its train—since in the subtle and luminous proto-matter everything was bound together and as one—the exteriorization or the "materialization" of all other earthly creatures, and thus also their "crystallization" in sensible, heavy, opaque and mortal matter.

Plato in his *Symposium* recalls the tradition that the human body, or even simply any living body, is like half a sphere; all our faculties and movements look and tend towards a lost center—which we feel as if "in front" of us—lost, but found again symbolically and indirectly, in sexual union. But the outcome is only a grievous renewal of the drama: a fresh entry of the spirit into matter. The opposite sex is only a symbol: the true center is hidden in ourselves, in the heart-intellect. The creature recognizes something of the lost center in his partner; the love which results from it is like a remote shadow of the love of God, and of the intrinsic beatitude of God; it is also a shadow of the knowledge which consumes forms as by fire and which unites and delivers.

The whole cosmogonic process is found again, in static mode, in man: we are made of matter, that is to say of sensible density and of "solidification," but at the center of our being is the supra-sensible and transcendent reality, which is at once infinitely fulminating and infinitely peaceful. To believe that matter is the "alpha" which gave to everything its beginning amounts to asserting that our body is the starting point of our soul, and that therefore the origin of our ego, our intelligence and our thoughts is in our bones, our muscles, our organs. In reality, if God is the "omega," He is of necessity also the "alpha," on pain of absurdity. The cosmos is a "message from God to Himself by Himself" as the Sufis would say, and God is "the First and the Last," and not the Last only. There is a sort of "emanation," but it is strictly discontinuous because of the transcendence of the Principle and the essential incommensurability of the degrees of reality; emanationism, on the contrary, is based on the idea of a continuity such as would not allow the Principle to remain unaf-

15 An expression used by René Guénon in speaking of the "realization of the supreme Identity." It is possible to consider deification as resembling—in the inverse direction—its antipodes, creation.

fected by manifestation. It has been said that the visible universe is an explosion and consequently a dispersion starting from a mysterious center; what is certain is that the total Universe, the greater part of which is invisible to us in principle and not solely de facto, describes some such movement—in an abstract or symbolical sense—and arrives finally at the deadpoint of its expansion; this point is determined, first by relativity in general and secondly by the initial possibilities of the cycle in question. The living being itself resembles a crystallized explosion, if one can put it in that way; it is as if the being had been turned to crystal by fear in the face of God.

*

Man, having shut himself off from access to Heaven and having several times repeated, within ever narrower limits, his initial fall, has ended by losing his intuition of everything that surpasses himself. He has thus sunk below his own true nature, for one cannot be fully man except by way of God, and the earth is beautiful only by virtue of its link with Heaven. Even when man retains belief, he forgets more and more what the ultimate demands of religion are; he is astonished at the calamities of this world, without its occurring to him that they may be acts of grace, since they rend, like death, the veil of earthly illusion, and thus allow man "to die before death," and so to conquer death.

Many people imagine that purgatory or hell are for those who have killed, stolen, lied, committed fornication and so on, and that it suffices to have abstained from these actions to merit Heaven. In reality, the soul is consigned to the flames for not having loved God, or for not having loved Him enough; this is understandable enough in the light of the supreme Law of the Bible: to love God with all our faculties and all our being. An absence of this love[16] does not necessarily involve murder or lying or some other transgression, but it does necessarily involve indifference;[17] and indifference, which is the most generally widespread of faults, is the very hallmark of the fall. It is possible for the indifferent[18] not to be criminals, but it is

---

16 It is not exclusively a question of a *bhakti,* of an affective and sacrificial way, but simply of the fact of preferring God to the world, whatever may be the mode of this preference; "love" in the Scriptures consequently embraces also the sapiential ways.

17 Fénelon was right to see in indifference the gravest of the soul's ills.

18 The *ghâfilûn* of the Koran.

impossible for them to be saints; it is they who go in by the "wide gate" and follow the "broad way," and it is of them that the book of Revelation says "So then because thou art lukewarm, and neither cold nor hot, I will spue thee out of my mouth" (Rev. 3:16). Indifference towards truth and towards God borders on presumption and is not free from hypocrisy; its seeming harmlessness is full of complacency and arrogance; in this state of soul, the individual is contented with himself, even if he accuses himself of minor faults and appears modest, which in fact commits him to nothing but on the contrary reinforces his illusion of being virtuous. It is this criterion of indifference that makes it possible for the "average man" to be "caught in the act," and for the most surreptitious and insidious of vices to be as it were taken by the throat, and for every man to have his poverty and distress proven to him; in short, it is indifference that is "original sin," or its most general manifestation.

Indifference is diametrically opposed to spiritual impassibility or to contempt of vanities, as well as to humility. True humility is to know that we can add nothing to God and that, even if we possessed all possible perfections and had accomplished the most extraordinary works, our disappearance would take nothing away from the Eternal.

Even believers themselves are for the most part too indifferent to feel concretely that God is not only "above" us, in "Heaven," but also "ahead" of us, at the end of the world, or even simply at the end of our own lives; that we are drawn through life by an inexorable force and that at the end of the course God awaits us; that the world will be submerged and swallowed up one day by an unimaginable irruption of the purely miraculous—unimaginable because surpassing all human experience and standards of measurement. Man cannot possibly draw on his past to bear witness to anything of the kind, any more than a may-fly can expatiate on the alternation of the seasons; the rising of the sun can in no way enter into the habitual sensations of a creature born at midnight whose life will last but a day; the sudden appearance of the orb of the sun, unforeseeable by reference to any analogous phenomenon that had occurred during the long hours of darkness, would seem like an unheard of apocalyptic prodigy. And it is thus that God will come. There will be nothing but this one advent, this one presence, and by it the world of experiences will be shattered.

*

In man stamped with the fall, not only has action priority over contemplation, but it even abolishes contemplation. Normally, the alternative ought not to be in evidence, contemplation being in its essential nature neither allied to action nor at enmity with it; but fallen man is precisely not "normal" man in the absolute sense. One could also say that in certain contexts there is harmony between contemplation and action whereas in other contexts there is opposition; but any such opposition is extrinsic and quite accidental. There is harmony in the sense that in principle nothing can be opposed to contemplation—this is the initial thesis of the Bhagavad Gîtâ—and there is opposition insofar as their respective planes differ; just as it is impossible to contemplate a nearby object and at the same time the distant landscape behind it, so too it is impossible—in this connection alone—to contemplate and to act at the same time.[19]

Fallen man is man led on by action and imprisoned by it, and that is why he is also sinful man; the moral alternative arises less from action than from the exclusivism of action, that is to say, from individualism with its illusion of being situated in a "territory" other than the "territory" of God; action becomes in a sense autonomous and totalitarian, whereas it ought to be fitted into a divine context, in a state of innocence wherein the separation of action from contemplation could not take place.

Fallen man is simultaneously squeezed and torn asunder by two pseudo-absolutes: the ponderous "I" and the dissipating "thing," the

---

19 This is what the tragedy of Hamlet expresses: facts and actions, and the exigencies of action were inescapable, but Shakespeare's hero saw through it all, he saw only principles or ideas; he plunged into things as into a morass; their very vanity, or their unreality, prevented him from acting, dissolved his action; he had before him, not this or that evil, but evil as such, and he broke himself against the inconsistency, the absurdity, the incomprehensibility of the world. Contemplation either removes action to a distance by causing the objects of action to disappear, or it renders action perfect by making God appear in the agent. The contemplativity of Hamlet had unmasked the world, but it was not yet fixed in God; it was as it were suspended between two planes of reality. In a certain sense, the drama of Hamlet is that of the *nox profunda:* it is also perhaps, in a more outward sense, the drama of the contemplative who is forced to action, but has no vocation for it; it is in any case a drama of profundity faced with the unintelligibility of the human comedy.

subject and the object, the ego and the world. As soon as he wakes up in the morning man remembers who he is; and straightway he thinks of one thing or another; between ego and object there is a link, which is usually action, so that a ternary is implicit in the phrase: "I—do—this" or, what amounts to the same thing: "I—want—this." Ego, act and thing are in effect three idols, three screens hiding the Absolute; the sage is he who puts the Absolute in the place of these three terms; it is God within him who is the transcendent and real Personality, and is hence the Principle of his "I."[20] His act is then the affirmation of God, in the widest sense, and his object is again God;[21] it is this that is realized, in the most direct way possible, by quintessential prayer[22] or concentration, which embraces, virtually or effectively, the whole of life and the whole world. In a more outward and more general sense, every man ought to see the three elements "subject," "act" and "object" in God, to the extent that he is capable of doing so through his gifts and through grace.

Fallen man is a fragmentary being, and therein lies a danger of deviation; for to be fragmentary is, strictly speaking, to lack equilibrium. In Hindu terms, one would say that primordial man, *hamsa,* was still without caste; the *brâhmana* however does not correspond exactly to the *hamsa,* he is only the uppermost fragment of the *hamsa,* otherwise he would by definition possess to the full the qualification of the warrior-king, the *kshatriya,* which is not the case; but every *Avatâra* is necessarily *hamsa,* and so is every "living liberated one," every *jîvan-mukta.*

A parenthesis may be permissible at this point. Mention has often been made elsewhere of the "naturally supernatural" transcendence of the Intellect; now one must not lose sight of the fact that this transcendence can act without impediment only on condition that it is framed by two supplementary elements, one human and the other divine, namely virtue and grace. "Virtue" in this sense is not equivalent to the natural qualities which of necessity accom-

20 "The Christ in me," as St. Paul would say.

21 This corresponds to the Sufi ternary "the one who invokes, the invocation, the One who is invoked" *(dhâkir, dhikr, Madhkûr).*

22 Such as the *japa* of the Hindus, the *dhikr* of the mystics of Islam, or the Jesus Prayer of the Hesychasts.

pany a high degree of intellectuality and contemplativity, it is a conscious and permanent striving after perfection, and perfection is essentially self-effacement, generosity and love of truth; "grace" in this sense is the divine aid which man must implore and without which he can do nothing, whatever his gifts; for a gift serves no purpose if it be not blessed by God.[23] The Intellect is infallible in itself, but this does not prevent the human receptacle from being subject to contingencies which, though they cannot modify the intrinsic nature of intelligence, can nonetheless be opposed to its full actualization and to the purity of its radiance.

With that in mind, let us return to the problem of action. The process of the fall, and even its results as well, are repeated on a reduced scale in every outward or inward act which is contrary to the universal harmony, or to any reflection of that harmony, such as a sacred Law. The man who has sinned has, in the first place, allowed himself to be seduced, and in the second place has ceased to be what he was before; he is as it were branded by the sin, and he is so of necessity, since every act must bear its fruit; every sin is a fall, and that being so it is also "the fall." Within the general conception of "sin," distinctions must be made between a "relative" or extrinsic sin, an "absolute"[24] or intrinsic sin, and a sin of intention. Sin is "relative" when it contravenes only some specific system of morality—such as polygamy in the case of Christians or wine in the case of Muslims—but then, by the very fact of this contravention, it amounts in effect for those concerned to "absolute sin," as is proven by the sanctions for the hereafter pronounced by the respective Revelations; nonetheless, certain "relative sins" can become legitimate—within the very framework of the Law which they contravene—under certain special circumstances; such, for example, is the case with killing in war. Sin is "absolute" or intrinsic when it is contrary to every code of morality and is excluded in all circumstances, like blasphemy, or contempt for truth. As for the sin of

---

23 In certain disciplines it is the *guru* who acts on behalf of God; the result in practice is the same, if account be taken of the conditions—and the imponderables—of the spiritual climate in question.

24 Needless to say, the word "absolute" when used in connection with sin is synonymous with "mortal"; it can have no more than a purely provisional and indicative function when, as in the present case, the ground it covers falls entirely within the actual framework of contingency.

intention, it is outwardly in conformity with a particular code or with all codes of morality, but inwardly opposed to the divine Nature, like hypocrisy for example. "Sin" is thus defined as an act which, firstly, is opposed to the divine Nature in one or another of its forms or modes (the reference here is to the Divine Qualities and the intrinsic virtues which reflect them) and which, secondly, engenders in principle posthumous suffering; it does so "in principle," but not always in fact, for repentance and positive acts on the one hand and the divine Mercy on the other efface sins, or can efface them. A "code of morality" in this sense is a sacred Legislation insofar as it ordains certain actions and prohibits certain others, independently of the depth or subtlety with which a particular doctrine may define its laws in other respects. This reservation is necessary because India and the Far East have conceptions of "transgression" and "Law" more finely shaded than those of the Semitic and European West, in the sense that, broadly speaking, in the East the compensatory virtue of knowledge is taken into account; it is "the lustral water without equal," as the Hindus say; and in the sense that intention plays a much more important part than most Westerners imagine, so that it can even happen, for example, that a *guru* should ordain, provisionally and with a view to some particular operation of spiritual alchemy,[25] actions which, while harming no one, are contrary to the Law;[26] but nonetheless a Legislation does comprise a code of morality, and man as such is so made that he distinguishes, rightly or wrongly, between a "good" and an "evil," that is to say his perspective is of necessity fragmentary and analytical. Moreover, the statement that certain acts are opposed to the "divine Nature" is made with the reservation that metaphysically nothing can be opposed to that Nature; Islam expresses this when it affirms that nothing can be separated from the divine Will, not even sin;[27] such ideas are not unconnected with

---

25 Islam is not ignorant of this point of view, witness the Koranic story of the mysterious sage scandalizing his disciples by actions with a secret intention, but externally illegal.

26 Or more precisely to the "prescriptions," such as exist in Hinduism and, in the West, especially in Judaism; there can be no question of infringements such as would seriously harm the collectivity.

27 Christianity also admits this idea because it could not do otherwise, but puts less emphasis on it.

25

non-Semitic perspectives, which always insist strongly on the relativity of phenomena, and on the variability of definitions to accord with different aspects of truth.

It is this essential and as it were supra-formal conception of sin which explains how in a tradition remaining "archaic" and therefore to a large extent "inarticulate," like Shinto for example, an elaborated doctrine of sin is absent; the rules of purity are the supports of a primordial synthetic virtue, superior to actions and considered as conferring on them a spiritual quality. Whereas Semitic morals start from action—outside esoterism at any rate—and seem to confine virtue to the realm of action and even to define it in terms of action, the moral code of Shinto and analogous codes[28] take an interior and global virtue as their starting-point and do not see acts as independent and self-contained crystallizations; it is only a posteriori and as a consequence of the "externalizing" influence of time, that the need for a more analytical code of morality could make itself felt.

Sin, as has been said, retraces the fall. But sin is not the only thing that retraces it in the realm of human attitudes and activities; there are also factors much more subtle and at the same time less serious, which intervene in a well-regulated life, and are connected with the kind of spiritual influences the Arabs call *barakah;* these factors become perhaps increasingly important as the spiritual aim becomes higher. They are connected, on the most diverse levels, with the choice of things or of situations; with the intuition of the spiritual quality of forms, gestures, morally neutral actions; their domain is connected with symbolism, aesthetics, with the significance of materials, proportions, movements, in short with everything which in a sacred art, a liturgy, a protocol, has meaning and importance. From a certain point of view, all this might seem negligible, but it is no longer at all so when one thinks of the "handling of spiritual influences"—if this expression be allowable—and when one takes account of the fact that there are forms which attract the presences of angels while there are others which repel them; in the same line of thought, one can say that, in addition to obligation, there is also a kind of courtesy towards Heaven. Things have their

---

28 One might well wonder whether "morality" is really the right word here, but that is a matter of terminology which is of little importance when the context admits of no misunderstanding.

cosmic relationships and their perfumes, and all things ought to retain something of a recollection of Paradise; life must be lived according to the forms and rhythms of primordial innocence and not according to those of the fall. To act according to *barakah* is to act in conformity with a kind of "divine aesthetic"; it is an outward application of the "discerning of spirits" or of the "science of humors" *(ilm al-khawâtir* in Arabic) as well as also of a geometry and of a music at once sacred and universal. Everything has a meaning and everything signifies something; to feel this and to conform to it is to avoid many errors that reason could not by itself prevent. Sacred art, which depends on this science of *barakah,* enfolds and penetrates the whole of human existence in traditional civilizations, and even constitutes all that is understood in our days by "culture," at least so far as those civilizations are concerned; but without this science of "benedictions" sacred art and all the forms of courtesy would remain unintelligible and would have no sense or value whatever.

What matters to the man who is virtually liberated from the fall is to remain in holy infancy. In a certain sense, Adam and Eve were "children" before the fall and became "adult" only through it and after it; the adult age in fact reflects the reign of the fall; old age, in which the passions are silenced, once again draws near to infancy and to Paradise, at any rate in normal spiritual conditions. The innocence and confidence of the very young must be combined with the detachment and resignation of the old; the two ages rejoin one another in contemplativity, and then in nearness to God: infancy is "still" close to Him, old age is so "already." The child can find his happiness in a flower, and so can the old man; the extremes meet, and life's spiral becomes a circle as its ends are brought together once more in the divine Mercy.

2

Sacred and Profane Science

René Guénon

In civilizations possessing a traditional character intellectual intuition occupies the position of a principle to which everything else can be referred; in other words it is the purely metaphysical doctrine that constitutes the essential, everything else being linked to it in the form either of consequences or of applications to the various orders of contingent reality. This is especially true of social institutions; but it is also true of the sciences, of those branches of knowledge, that is to say, which are concerned with the sphere of the relative and can only be regarded, in such civilizations, as dependencies and, as it were, prolongations or reflections of absolute and principial knowledge.[1] It is in this manner that the proper hierarchy is everywhere and always preserved: the relative is not in any way treated as nonexistent, which would be senseless; it is duly taken into consideration, but it is placed in its proper posi-

---

1 *Editor's note:* "The whole existence of the peoples of antiquity, and of traditional peoples in general, is dominated by two presiding ideas, the idea of Center and the idea of Origin. In the spatial world we live in, every value is related back in one way or another to a sacred Center, to the place where Heaven has touched the earth; in every human world there is a place where God has manifested Himself to spread His grace therein. Similarly for the Origin, the quasi-timeless moment when Heaven was near and when terrestrial things were still half-celestial; but the Origin is also, in the case of civilizations having a historical founder, the time when God spoke, thereby renewing the primordial alliance for the branch of humanity concerned. To conform to tradition is to keep faith with the Origin, and for that very reason it is also to be situated at the Center; it is to dwell in the primordial Purity and in the universal Norm. Everything in the behavior of ancient and traditional peoples can be explained, directly or indirectly, by reference to these two ideas, which are like landmarks in the measureless and perilous world of forms and of change." Frithjof Schuon, *Light on the Ancient Worlds* (World Wisdom, Bloomington, 1984), p. 7.

tion, which cannot be other than a secondary and subordinate one; and within this sphere of the relative itself there are many different degrees, depending upon whether the subject under consideration lies closer to or farther away from the realm of principles.

Thus, as far as the sciences are concerned, there are two radically different and even incompatible conceptions, which may be referred to respectively as the traditional and the modern conceptions; we have often had occasion to allude to those "traditional sciences" which existed in Antiquity and in the Middle Ages and which still exist in the East today,[2] although the very notion of any such thing has become completely foreign to the Occidentals. It should be added that every civilization has possessed "traditional sciences" of a particular sort peculiar to itself, the reason being that where sciences are concerned one is no longer in the sphere of universal principles, which is the province of pure metaphysics alone, but in the realm of adaptations; in this realm, for the very reason that it is a contingent one, account has to be taken of the whole aggregate of conditions, mental and otherwise, which belong to any given people and one may even say, to any given period in the existence of a people, since there are periods when "readaptations" become necessary. These readaptations are no more than changes of form, not affecting the essence of the tradition in any way; as far as metaphysical doctrine is concerned only the expression can be modified, in a manner more or less comparable to translation from one language into another; though the forms may be various which it assumes for the sake of expressing itself, insofar as such expression is possible, metaphysics remains one, just as truth is but one.

When one passes, however, to the realm of applications the case is naturally altered: with the sciences, as with social institutions, one enters the world of form and multiplicity; it is on this account that differences of form may really be said to constitute different sciences, even when the object of study remains at least partially the

2 *Editor's note:* This article was first published in 1927. Since that time the trend that René Guénon so well described in his monumental work *The Reign of Quantity and the Signs of the Times* (Sophia Perennis, Ghent, New York, 2001), has continued towards the final stages of the Kali Yuga with the ensuing destruction of the traditional worlds and their respective cultural frameworks, including the traditional sciences and crafts. This element is analyzed in detail by Frithjof Schuon in his *The Eye of the Heart* (World Wisdom, Bloomington, 1997), Chapter 8.

same. Logicians are accustomed to regard a science as entirely defined by its object, but this is an over-simplified view; the standpoint from which the object is envisaged must also enter into the definition of a science. The number of possible sciences is indefinite; it can well happen that several sciences will study the same things, but under such different aspects and therefore by such different methods and with such different intentions, that they are nonetheless in reality quite distinct sciences. This is especially liable to happen with "traditional sciences" belonging to different civilizations; sciences, that is to say, which, although mutually comparable, nevertheless cannot always be assimilated to one another and often could not correctly be described by the same name. It goes without saying that the difference is still more marked if, instead of making a comparison between traditional sciences, which do at least all possess the same character fundamentally, one tries to compare these sciences in a general way with science as conceived by the modern world; at first sight it might sometimes appear that the object of study was the same in either case and yet the knowledge of it which the two kinds of science provide differs so widely that one hesitates, upon closer examination, to continue regarding them as the same, even in a partial sense.

A few examples may serve to make our meaning clearer; and to begin with we will take a very general one, namely that of "physics," as understood by the ancients and by the moderns respectively; in this case moreover it is not necessary to look beyond the western world in order to observe the profound difference separating the two conceptions. The term "physics" in its original and etymological sense meant nothing more nor less than the "science of nature" without qualification of any kind; it is therefore a science which deals with the most general laws of "becoming" ("nature" and "becoming" being synonymous fundamentally), and it was in this way that the Greeks, and notably Aristotle, understood this science; if more specialized sciences happen to exist relating to the same order, they can amount to no more than "specifications" of physics, dealing with some more narrowly defined sphere or other. Already therefore there is something rather significant about the deviation of meaning to which the moderns have subjected the word "physics" by reserving it exclusively to describe one particular science among many others, all of which are equally natural sciences; this fact is closely connected with that process of subdivision that we have remarked upon as a characteristic of modern science, a form of "specialization" bred of the analytical frame of mind and carried to such lengths as to render the conception of a single science treating nature as one whole well-nigh inconceivable to anyone who has undergone its influence. The inconveniences resulting from this specialization, and above all the narrowness of outlook it engenders, have not passed altogether unnoticed; but it would seem that those very people who are most clearly aware of the fact have resigned themselves to it nevertheless as a necessary evil resulting from the vast accumulation of detailed knowledge which no one man could ever hope to grasp; on the one hand they have not understood that such detailed knowledge lacks significance in itself and is not worth the sacrifice of a synthetic knowledge belonging to a much higher order even though still dealing with the relative; and on the other hand they have failed to see that the impossibility of unifying the multiplicity of this detailed knowledge is a consequence of their own reluctance to relate it to a higher principle; it is due, that is to say, to a persistence in working from the bottom upwards and from externals, whereas the very opposite process is called for if one wishes to possess sciences endowed with real speculative value.[3]

If, instead of comparing the physics of the ancients with what the moderns understand by the term, one were to compare it with the whole aggregate of natural sciences as at present constituted—and that is what ought really to correspond to the ancient physics—the first point of difference to note would be the subdivision into "specialities" which are so to speak foreign to one another. This is however only the most external aspect of the question and it must not be supposed that by combining all these special sciences one

3 *Editor's note:* "In all this wish [of modern science] to accumulate knowledge of relative things, the metaphysical dimension—which alone takes us out of the vicious circle of the phenomenal and the absurd—is expressly put aside; it is as if a man were endowed with all possible faculties of perception minus intelligence; or again, it is as if one believed that an animal endowed with sight were more capable than a blind man of understanding the mysteries of the world. The science of our time knows how to measure galaxies and split atoms, but it is incapable of the least investigation beyond the sensible world, so much so that outside its self-imposed but unrecognized limits it remains more ignorant than the most rudimentary magic." Frithjof Schuon, *Treasures of Buddhism* (World Wisdom, Bloomington, 1993), p. 42.

would arrive at the equivalent of the ancient physics. The fact of the matter is that the point of view is completely alien and it is here that the essential difference arises between the two conceptions referred to above; the traditional conception, as we have already remarked, links all the sciences to the principles of which they become particular applications, and it is precisely this connection which the modern conception fails to admit. For Aristotle, physics came "second" in relation to metaphysics, it was dependent upon metaphysics that is to say, and was really only an application to the province of nature of principles which are superior to nature and are reflected in her laws; and the same can be said of mediaeval cosmology. The modern conception, on the other hand, claims to make the sciences independent by repudiating everything that transcends them, or at least by declaring it "unknowable" and refusing to take it into account, which amounts to ignoring it in practice; this negation existed as a fact for a long time before people thought of erecting it into a systematic theory under such names as "positivism" and "agnosticism," for it may truly be said to lie at the root of modern science as a whole. It is only in the nineteenth century however that one finds men glorying in their ignorance (since to call oneself an "agnostic" amounts to nothing else), and claiming to deny others all knowledge of the things they themselves are ignorant of, and that stage marked a further step in the intellectual decline of the West.

In seeking completely to sever the connection between the sciences and any higher principles, on the pretext of safeguarding their independence, the modern conception robs them of all deeper meaning and even of any real interest from the point of view of knowledge, and it can only lead them down a blind alley, imprisoning them, as it does, within an incurably limited realm.[4] Moreover the development which goes on inside that realm is not a

---

4 It should be noted that something similar has occurred in the social order, where the moderns claim to separate the temporal from the spiritual. It is not a question of denying the fact that the two are distinct, since they refer effectively to different realms, just as in the case of metaphysics and the sciences. What is overlooked, however, thanks to an inherent error of the analytical approach, is that distinction does not mean complete separation. In this way, the temporal power forfeits its legitimacy, and the same could be said of the sciences, in the intellectual order.

deepening of knowledge, as is commonly supposed; on the contrary, the information so gained remains superficial and consists merely in that dispersion in detail that we have already alluded to, in an analysis as barren as it is laborious and which can be pursued indefinitely without advancing a single step further in the direction of true knowledge. Furthermore it is not for its own sake that Westerners in general cultivate science as they understand it; their primary aim is not knowledge, even of an inferior order, but practical applications, as may be inferred from the ease with which the majority of our contemporaries confuse science and industry, so that by many the engineer is looked upon as a typical man of science; but this is connected with another question that we shall have to go into more fully later on.

In assuming its modern form science has not only lost in depth, but also, one might say, in solidity, since attachment to the principles enabled it to participate in their immutability to the full extent that the nature of its subject matter allowed; once shut off exclusively in the realm of change, however, it cannot hope to achieve any kind of stability, nor to find any solid basis on which to build; no longer starting out from any certainty, it finds itself reduced to probabilities and approximations, or to purely hypothetical constructions which are merely the product of individual fantasy. Furthermore, even if modern science accidentally happens to arrive, by a very roundabout route, at certain results which appear to agree with some of the data of the ancient traditional sciences, it would be the greatest mistake to look upon those results as confirming the data in question, which stand in no need of any such confirmation; and it would be a waste of time to try and reconcile such totally different points of view, or to establish a concordance with hypothetical theories which may be completely discredited in a few years time.[5] So far as modern science is concerned these conclusions cannot but partake of the nature of hypotheses, whereas they amounted to something quite different for the "traditional sciences," presenting themselves as the unquestionable consequences

---

5 The same observation applies, from the religious point of view, to a certain "apologetic" which claims to establish an agreement with the results of modern science, an utterly illusory task and one constantly needing to be started anew, involving the grave danger of appearing to bind up religion with changing and ephemeral conceptions of which it should remain totally independent.

of truths known intuitively, and therefore infallibly, within the metaphysical order.[6] Moreover it is a peculiar delusion, typical of modern "experimentalism," to suppose that a theory can be proved by facts, whereas really the same facts can always be equally well explained by a variety of different theories; and certain of the pioneers of the experimental method, such as Claude Bernard, have themselves recognized that they could interpret facts only with the help of "preconceived ideas," apart from which they would remain "bare facts," devoid of significance or scientific value.

While speaking of "experimentalism" the opportunity may be taken to reply to a question which is sometimes raised in this connection, and which is as follows: why have the experimental sciences received a development in the modern civilization such as they never received at the hands of any other civilization before? The reason is that they confine their attention to things of the senses and to the world of matter, and also that they lend themselves readily to the most immediate practical applications; their development, going hand in hand with what may well be termed the "superstition of facts," is thus quite in accordance with the specifically modern tendencies, whereas preceding ages would, on the contrary, have been unable to find sufficient inducements for becoming absorbed in this direction to the extent of neglecting the higher orders of knowledge. It should be clearly understood that, according to our view, there is no question of maintaining that any kind of knowledge, however inferior, is illegitimate in itself; what is not legitimate is simply the abuse which occurs when subjects of this kind absorb the whole of human activity, as is the case today. One might even conceive of a normal civilization where there were experimental sciences attached, like the other sciences, to the principles and thus provided with a real speculative value; in point of fact, if no such instance seems to have occurred, that is because attention was turned for preference in other directions, and also because, even when it was a question of studying the sensible world insofar as it might appear interesting to do so, traditional data

---

6 It would be easy to give examples of this: we will mention only, as being one of the most striking, the different nature of the conceptions of ether to be found in Hindu cosmology and in modern physics.

would have made it possible to undertake this study more advantageously by other methods and from a different point of view.

We remarked above that one of the characteristics of the present time is the exploitation of all those things that had hitherto been neglected as not possessing sufficient importance for men to devote their attention to them, but which nevertheless had also to be developed before the end of the present cycle, since they too have their place among the possibilities destined to be manifested therein; such in particular is the case of the experimental sciences which have come into existence during the course of recent centuries. There are even certain modern sciences which actually amount, in the most literal sense of the word to "residues" of ancient sciences that are no longer understood:[7] it is the most inferior elements of these latter sciences which, through being isolated and detached from all the rest during a period of decadence, became grossly materialized and then served as the starting point for quite a different development along lines conforming with modern tendencies, in such a way as to lead to the formation of sciences no longer having anything in common with those that had preceded them. Thus for instance it is incorrect to maintain, as is generally supposed, that astrology and alchemy have respectively become modern astronomy and chemistry, even though this view contains a certain degree of truth from the purely historical angle, just so much in fact as is apparent from what we have said above: if the latter sciences have indeed issued from the former in a certain sense, it is not as the result of "evolution" or "progress," as is commonly asserted, but, on the contrary, by a process of degeneration; and this is a point which calls for further explanation.

In the first place it should be noted that the attribution of a separate meaning to the terms "astrology" and "astronomy" is of relatively recent origin; among the Greeks both words were employed, without distinguishing between them, in order to denote the whole of the field now divided up between the two terms. It would seem then, at first sight, as if this were but another instance of that division introduced for the sake of "specialization" between what were

7 *Translator's note.* It is worthy of notice that the Tibetan name for the Kali Yuga is, literally, "the age of impure residues." Its final phase is likewise described as "the time when impurities grow more and more."

originally only parts of a single science; but what is peculiar in the present case is that, whereas one of the parts, that namely which represented the more material side of the science in question, underwent an independent development, the other part, on the contrary, disappeared altogether. So true is this that it is not even known any longer what ancient astrology amounted to, and even those who have attempted to reconstruct it never achieve more than a counterfeit of it; either they attempt to turn it into the equivalent of a modern experimental science and have recourse to statistics and the calculation of probabilities, in consequence of the adoption of a point of view that could not possibly have existed for either the ancient or the mediaeval worlds, or else they direct their attention exclusively to the restoration of an "art of divination" which amounted to no more than a perversion of astrology in its decline and which could be regarded at most as a very inferior application, scarcely worthy of serious consideration, as can still be observed in the attitude shown towards it in the East today.

The case of chemistry is perhaps even clearer and more typical; and, as regards the ignorance of the moderns about the true nature of alchemy, it is at least as great as in the case of astrology. Genuine alchemy was essentially a science belonging to the cosmological order, and at the same time it was also applicable to the human order, by virtue of the analogy between the "macrocosm" and the "microcosm"; furthermore, it was constituted particularly with a view to allowing of a transposition into the purely spiritual realm, which lent a symbolical value and a higher significance to its teaching, placing it among the most complete types of "traditional sciences." It is not from this alchemy, with which, as a matter of fact, it has nothing in common, that modern chemistry has sprung; modern chemistry is a corruption and, in the strictest sense of the word, a deviation having its origin, perhaps as early as the Middle Ages, in a lack of understanding on the part of persons who, from incapacity to penetrate the true meaning of the symbols used, took everything literally and launched out into a more or less confused experimentalism on the supposition that alchemy was purely and simply a question of material manipulations. These people, sarcastically referred to by the alchemists as "blowers" and "charcoal burners" were the real forerunners of the chemists of today; and this illustrates how modern science came to be built up from the remnants of ancient sciences, with materials which had been rejected and abandoned to the ignorant and the "profane." Let it be added that the so-called restorers of alchemy, of whom there are a certain number to be found in the contemporary world, are for their part merely prolonging this very deviation, and their researches are as far removed from traditional alchemy as are those of present day astrologers from ancient astrology, and it is for this reason that one is justified in declaring that the "traditional sciences" of the West really are lost for the modern world.

We will confine ourselves to these few examples, although it would be an easy matter to supply a number of others chosen from various different spheres and revealing a similar degeneration everywhere. It could be shown in this way that psychology as understood today, the study, that is to say, of mental phenomena as such, is a natural product of Anglo-Saxon empiricism and of the attitude of mind of the eighteenth century, and that the point of view to which it corresponds was so secondary in the eyes of the ancients that, even if it had happened to be taken into consideration incidentally, it could under no circumstances have been erected into a special science; whatever of value may be contained in it was to be found transformed and assimilated, as far as they were concerned, in accordance with higher points of view. In quite a different sphere it might also be shown that modern mathematics represents no more than the outer crust, so to speak, or the exoteric side, of Pythagorean mathematics; the ancient conception of numbers has even become quite unintelligible to the moderns, since, in that case as well, the superior portion of the science, that which, along with its traditional character, gave it genuine intellectual value, has disappeared completely, and the case of mathematics is very similar to that of astrology. But to pass all the sciences in review, one after another, would be tedious; enough has been said to explain the nature of the change to which the modern sciences owe their birth and which is the very opposite of a "progress" amounting rather to a veritable regression of intelligence; and we will now return to questions of a general order concerning the parts played by "traditional" and modern sciences respectively and the profound differences which exist between them as to their true aims.

According to the traditional conception a science is interesting not so much for its own sake as for its being as it were a prolongation or secondary branch of the doctrine, of which the essential part is constituted, as we have seen, by pure metaphysics.[8] Actually, if every science is certainly legitimate, so long as it does not overstep the position that properly belongs to it in virtue of its own nature, it will nevertheless be easily understood that, for anyone possessing knowledge of a higher order, the lower forms of knowledge inevitably lose a great deal of their interest; whatever interest they do retain will only be as a function, so to speak, of the principial knowledge, that is to say insofar as, on the one hand, they reflect this knowledge in such and such a contingent sphere, or, on the other hand, insofar as they are capable of leading up to that same principial knowledge, which, in such a case, must never be lost sight of or sacrificed to more or less accidental considerations. These are the two complementary functions that properly belong to the "traditional sciences": on the one hand, as applications of the doctrine, they allow of linking up all the different orders of reality one to another and of integrating them in the unity of the total synthesis; on the other hand, they constitute, for some people at least, and in accordance with their own particular aptitudes, a preparation for a higher type of knowledge and a kind of pathway leading towards it, while from their hierarchical arrangement, according to the levels of existence to which they relate, they form as it were so many rungs of a ladder with the aid of which it is possible to raise oneself to the heights of pure intellectuality.[9] It is only too easy to see that the modern sciences cannot in any way fulfill either the one or the other of these twin purposes; it is for this reason that they cannot amount to anything but "profane science," whereas the traditional sciences, owing to their link with the metaphysical principles, are effectively incorporated in "sacred science."

The twofold purpose that we have just pointed out does not moreover imply either a contradiction or a vicious circle, though

8 This is expressed, for example, in a title such as *upaveda,* used in India for certain traditional sciences and showing their subordination to the *Veda,* that is to say to sacred knowledge.

9 In our study, *L'Esotérisme de Dante* (The Esoterism of Dante [New York: Sophia Perennis et Universalis, 1996]. Ed.), we spoke of the symbolism of the ladder, the rungs of which correspond, in various traditions, to certain sciences and, at the same time, to states of the being; this necessarily implies that these sciences were not regarded in a merely "profane" manner, as in the modern world, but allowed of a transposition which bestowed on them a real initiatory significance.

superficially it might appear to do so; and this is also a point that requires explaining. It might be described as a question of two points of view, the one descending and the other ascending, or the one corresponding to an unfolding of knowledge, starting from the principles and proceeding towards ever more distant applications, and the other corresponding to a gradual acquisition of that same knowledge, proceeding from the lower to the higher, or, if preferred, from the outer to the inner. It is not therefore a matter of knowing whether the sciences ought to be constituted from below upwards or from above downwards, or whether it is necessary for their existence to take cognizance of principles or of the sensible world as their starting point; this question, which may arise from the standpoint of "profane" philosophy and seems indeed to have arisen more or less explicitly among the Greeks, does not exist at all for "sacred science" which cannot start out from anything except the universal principles; the reason why such a question does not apply in this case is that the prime factor here is intellectual intuition, which is the most direct of all forms of knowledge as well as the highest, and is absolutely independent of the exercise of any faculty belonging to the sensible or even to the rational order. Sciences can only be validly constituted as "sacred sciences" by those who, above all else, are in full possession of principial knowledge and who alone are qualified, on that very account, to carry out, in conformity with the strictest traditional orthodoxy, all the various adaptations necessitated by circumstances of time and place. Once the sciences have been constituted in this manner, however, the teaching of them may follow an inverse order; they will serve as "illustrations" of the pure doctrine, so to speak, which they are able to render more easily accessible to certain types of mind; and from the fact that they deal with the world of multiplicity they are adapted, through the almost indefinite variety of their points of view, to the equally wide variety of individual aptitudes found among those types of mind whose horizon is still confined to that same world of multiplicity; the possible paths leading to knowledge may be extremely varied at the lowest levels, but they will converge more and more as the higher degrees are reached. This does not mean to say that any of these preparatory degrees are absolutely necessary, since they amount to no more than contingent means and enjoy no common measure with the goal to be attained; it can

even happen that some among those in whom the tendency to contemplation predominates will arrive in a single leap at true intellectual intuition without the aid of any such means;[10] but these are
more or less exceptional cases and for the generality of men it is a
matter of convenience, if one may so put it, amounting to a practical necessity, that they should proceed upwards by gradual stages.
To make the point clearer one can also make use of the traditional
symbol of the "cosmic wheel": the circumference only exists really
in virtue of the center; but the beings who find themselves at the
circumference must necessarily start out from there and follow the
radius in order to reach the center. Furthermore, as a result of the
correspondence that exists between every order of reality, truths
belonging to a lower order can be taken as symbolical of those
belonging to a higher order, and thus act as "supports" for arriving
at a knowledge of the latter by the use of analogy; this it is which
endows a science with a superior or "anagogical" meaning deeper
than that which it possesses in itself, and bestows upon it the character of a genuine "sacred science."[11]

Every science, be it said, is capable of assuming this character,
whatever its subject matter, on the sole condition of being set up
and envisaged according to the traditional spirit; it is merely necessary to bear in mind the degrees of importance of the different sciences, depending upon the hierarchical position of the various
orders of reality dealt with; but, whatever the degree, their character and their function remain essentially the same in the traditional conception. What is true of all the sciences in this respect is
equally true of every art, inasmuch as an art can possess a genuinely
symbolical value which enables it to serve as a support for medita-
tion, and also because its rules like the laws which it is the object of
science to understand, are in their turn reflections and applications
of the fundamental principles; and thus it is that in every normal

10 This is why, according to the Hindu doctrine, the *Brâhmans* should keep their
minds constantly turned towards the supreme knowledge, whereas the *Kshatriyas*
should rather apply themselves to a study of the successive stages by which this is
gradually to be reached.

11 This is the purpose, for instance, of the astronomical symbolism so commonly used in the various traditional doctrines; and what we say here can serve to give
an idea of the true nature of ancient astrology.

civilization there are "traditional arts" which are no less lost to the modern West than the traditional sciences.[12] The truth is that there is really no such thing as a "profane realm" opposable in some way to a "sacred realm"; there is simply a "profane point of view" which is really nothing but the point of view of ignorance.[13] It is for this reason that profane science, as understood by the moderns that is to say, can fairly be described as "ignorant knowledge" as we have already remarked elsewhere: it is knowledge of an inferior order, remaining at the level of the lowest degree of reality and blind to everything that transcends it or to any aims loftier than its own, as well as to any principle capable of assuring it a legitimate place, however humble, among the various orders of knowledge as a whole; imprisoned irremediably within the relative and narrow field in which it has striven to proclaim itself independent, thereby of its own accord severing all connection with transcendent truth and supreme knowledge, it amounts to no more than an aimless and illusory form of knowledge, issuing out of nothing and leading nowhere.

This survey should suffice to make plain the deficiency of the modern world from a scientific standpoint and to show how that same science in which it takes such pride represents no more than a deviation and, as it were, a remnant of true science, which, in our eyes, can only be synonymous with what we have called "sacred" or "traditional" science. Modern science, arising out of an arbitrary limitation of knowledge within a certain particular order which is indeed the most inferior of all, namely that of material or sensible reality, has as a consequence forfeited all intellectual value, so long that is to say as one uses the word intellectuality in all the fullness of its true meaning and refuses to participate in the "rationalist" error, or to reject intellectual intuition, which amounts to the same thing. The source of this error as of a great many other modern errors,

---

12 The art of the mediaeval builders can be quoted as a particularly remarkable example of these traditional arts, whose practice, moreover, implied a real knowledge of the corresponding sciences.

13 To see the truth of this, it is sufficient to note facts such as the following: cosmogony, one of the most sacred of the sciences, a science which has its place in all the inspired books, including the Hebrew Bible, has become for the modern world a subject for completely "profane" hypotheses; the domain of the science is the same in both cases, but the point of view is utterly different.

and likewise the root of the entire deviation of science as outlined above, can be discovered in what may be called "individualism" an attitude of mind which is indistinguishable from the anti-traditional attitude itself and of which the numerous manifestations, apparent in every sphere, constitute one of the most important factors in the confusion of our time.[14]

14 *Editor's note:* "A science that penetrates the depths of the 'infinitely great' and of the 'infinitely small' on the physical plane, but denies other planes although it is they that reveal the sufficient reason of the nature we perceive and provide the key to it, such a science is a greater evil than ignorance pure and simple; it is in fact a 'counter-science,' and its ultimate effects cannot but be deadly. In other words, modern science is a totalitarian rationalism that eliminates both Revelation and Intellect, and at the same time a totalitarian materialism that is blind to the meta-physical relativity—and therewith also the impermanence—of the world. It does not know that the suprasensible, situated as it is beyond space and time, is the concrete principle of the world, and that it is consequently also the origin of that contingent and changeable coagulation that we call 'matter.' A science that is called 'exact' is in fact an 'intelligence without wisdom,' just as post-Scholastic philosophy is inversely a 'wisdom without intelligence'." Frithjof Schuon, *Light on the Ancient Worlds* (World Wisdom, Bloomington, Indiana, 1984), pp. 116-117.

3

Traditional Cosmology and the Modern World

Titus Burckhardt

1. The Cosmological Perspective

The seven "liberal arts" of the Middle Ages have as their object disciplines which modern man would automatically describe as "sciences" such as mathematics, astronomy, dialectic, and geometry. This medieval identification of science with art, wholly in conformity with the contemplative structure of the *Trivium* and *Quadrivium,* clearly indicates the fundamental nature of the cosmological perspective.

When modern historians look at traditional cosmology— whether this be the cosmological doctrines of ancient and oriental civilizations, or the cosmology of the medieval West—they generally see in it merely childish and groping attempts to explain the causation of phenomena. In so doing, they are guilty of an error in their way of looking at things which is analogous to the error of those who, with a "naturalistic" prejudice, judge medieval works of art according to the criteria of the "exact" observation of nature and of artistic "ingeniosity." Modern incomprehension of sacred art and contemplative cosmology thus arises from one and the same error; and this is not gainsaid by the fact that some scholars (often the very ones who look on oriental or medieval cosmology with a combination of pity and irony) pay homage to the arts in question and allow the artist the right to "exaggerate" some features of his natural models and to suppress others with a view to suggesting realities of a more inward nature. This tolerance only proves that, for modern man, artistic symbolism has no more than an individual, psychological—or even merely sentimental—bearing. Modern scholars are obviously unaware that the artistic choice of forms, when it pertains to inspired and regularly transmitted principles, is capable of tangibly conveying the permanent and inexhaustible possibilities of the Spirit, and that traditional art thus implies a "logic" in the universal sense of this term.[1] On the one hand, the modern mentality is blinded by its attachment to the sentimental aspects of art forms (and only too often reacts as a result of a very particular psychic heredity); on the other, its starting-point is the prejudice that artistic intuition and science belong to two radically different domains. If this were not so, one would have in all fairness to grant to cosmology what one seems to grant to art, namely the license to express itself by means of allusions and to use sensible forms as parables.

For modern man, however, any science becomes suspect if it leaves the plane of physically verifiable facts, and it loses its plausibility if it detaches itself from the type of reasoning that is completely reliant on, as it were, a plastic continuity of the mental faculty. As if it could possibly be justifiable to suppose that the whole cosmos were made so as to reflect merely the "material" or quantitative sides of the human imagination. Such an attitude moreover does not do justice to the full human reality. It represents more a mental limitation (resulting from an extremely unilateral and artificial activity) than a philosophical position, for all science, however relative or provisional it may be, presupposes a necessary correspondence between the order that is spontaneously inherent in the knowing mind and the compossibility of things, otherwise there would be no truth of any kind.[2] Now since the analogy between the macrocosm and the microcosm cannot be denied, and since it everywhere affirms principial unity—a unity that is like an axis in regard to which all things are ordered—it is impossible to see why the science (i.e. knowledge) of "nature" in the vastest possible sense of this term, should not reject the crutches of a more or less quantitative experience, and why any intellectual vision (possessed as it were of a "bird's eye view") should be immediately dismissed as a gratuitous hypothesis. But modern scientists have a veritable aver-

---

1 See Frithjof Schuon: *The Transcendent Unity of Religions,* (Theosophical Publishing House, Wheaton, Illinois, 1984) Chapter 4, "Concerning Forms in Art".

2 See René Guénon: *Introduction to the Study of the Hindu Doctrines* (Sophia Perennis, Ghent, New York, 2002), chapter on "*Nyaya*": " . . . if the idea, to the extent that it is true and adequate, shares in the nature of the thing, it is because, conversely, the thing itself also shares in the nature of the idea."

sion to anything that goes beyond the allegedly down-to-earth nature of "exact science." In their eyes, to have recourse to the poetical quality of a doctrine, is to discredit that doctrine as science. This heavy "scientistic" distrust of the grandeur and beauty of a given conception shows a total incomprehension of the nature of primordial art and of the nature of things.

Traditional cosmology always comprises an aspect of "art" in the primordial sense of this word: when science goes beyond the horizon of the corporeal world or when the traditional cosmologist gives his attention only to the manifestations, within this very world, of transcendent qualities, it becomes impossible to "record" the object of knowledge as one records the contours and details of a sensory phenomenon. We are not saying that the intellection of realities higher than the corporeal world is imperfect; we are referring only to its mental and verbal "fixation." Whatever can be conveyed of these perceptions of reality is inevitably in the form of speculative keys, which are an aid to rediscovering the "synthetic" vision in question. The proper application of these "keys" to the endless multiplicity of the faces of the cosmos is dependent on what may indeed be called an art, in the sense that it presupposes a certain spiritual realization or at least a mastery of certain "conceptual dimensions."[3]

As for modern science, not only is it restricted, in its study of nature, to only one of its planes of existence (whence its "horizontal" dispersion contrary to the contemplative spirit); it also dissects as far as possible the contents of nature, as if the more to emphasize the "autonomous materiality" of things; and this fragmentation—both theoretical and technological—of reality is radically opposed to the nature of art; for art is nothing without fullness in unity, without rhythm, without proportion.

In other words, modern science is ugly, with an ugliness that has finished by taking possession of the very notion of "reality"[4] and by arrogating to itself the prestige of the "objective" judgement of

---

3 An example of such a speculative "key" is the diagram of a horoscope. This symbolically summarizes all the relationships between a human microcosm and the macrocosm. The interpretation of a horoscope comprises innumerable applications, which, however, can only be properly divined by virtue of the unique "form" of the being, a form which the horoscope both reveals and veils.

4 Whence the use, in modern aesthetics, of the term "realism."

things,[5] whence the irony of modern men with regard to whatever, in the traditional sciences, may reveal an aspect of artless beauty. Conversely, the ugliness of modern science deprives it of any value from the point of view of the contemplative and inspired sciences, for the central object of these sciences is the unicity of everything that exists, a unicity that modern scientists cannot in fact deny—since everything implicitly affirms it—but which it nevertheless, by its dissecting approach, prevents one from "tasting."

2. Traditional Cosmology and Modern Science

(I) COSMOLOGIA PERENNIS

In what follows attention will be drawn to certain fissures in modern science, and these will be judged by means of the criteria provided by cosmology in the traditional sense of this term. We know that the Greek word *cosmos* means "order" implying the ideas of unity and totality. Cosmology is thus the science of the world inasmuch as this reflects its unique cause, Being. This reflection of the uncreated in the created necessarily presents itself under diverse aspects, and even under an indefinite variety of aspects, each of which has about it something whole and total, so that there are a multiplicity of visions of the cosmos, all equally possible and legitimate and springing from the same universal and immutable principles.

These principles, by reason of their very universality, are essentially inherent in human intelligence at its most profound; but this pure intellect only becomes "disengaged" generally speaking and for the man who is predisposed thereto, with the aid of supernatural elements that an authentic and complete spiritual tradition alone can supply. This means that all genuine cosmology is attached to a divine revelation, even if the object it considers and the mode of its expression apparently lie outside the message that this revelation brings.

---

5 For the overwhelming majority of moderns, the signs and characteristics of science are complex pieces of apparatus, endless reportings, a "clinical" approach, etc.

Such is the case, for instance, with Christian cosmology, whose origin at first sight appears somewhat heterogeneous, since on the one hand it refers to the Biblical account of creation, while on the other hand it bases itself on the heritage of the Greek cosmologists; if there seems to be a certain eclecticism here, it should be stressed that this is providential, since the two sources in question complement one another in a harmonious way, the first being presented in the form of a myth and the other under the form of a doctrine expressed in more or less rational terms, and thus neutral from the point of view of symbolism and of a spiritual perspective.

Moreover, there can only be a question of syncretism where there is a mixture, and hence a confusion, of planes and modes of expression. The Biblical myth of creation and Greek cosmology do not present any formally incompatible perspectives, nor do they duplicate one another, as would be the case, for example, if one attempted to mingle Buddhist cosmology with the figurative teaching of the Bible. The Biblical myth assumes the form of a drama, a divine action that seems to unfold in time, distinguishing the principial and the relative by a "before" and an "after." Greek cosmology, for its part, corresponds to an essentially static vision of things; it depicts the structure of the world, such as it is "now and always," as a hierarchy of degrees of existence, of which the lower are conditioned by time, space, and number, while the higher are situated beyond temporal succession and spatial or other limits. This doctrine thus presents itself quite naturally—and providentially—as a scientific commentary on the scriptural symbolism.

The Biblical myth is revealed, but Greek cosmology is likewise not of purely human origin; even with Aristotle, that distant founder of Western rationalism, certain basic ideas, like his distinction between form *(eidos)* and matter *(hyle)*, for example, undoubtedly spring from a knowledge that is supra-rational, and therefore timeless and sacred. Aristotle translates this wisdom into a homogeneous dialectic, and his dialectic is valid because the law inherent in thought reflects in its own way the law of existence. At the same time, he demonstrates reality only to the extent that it can be logically defined. Plato and Plotinus go much further; they transcend the "objectified" cosmology of Aristotle, and restore to symbolism all its supra-rational significance. Christian cosmology borrowed the analytical thought of Aristotle, but it was from Plato that it derived the doctrine of archetypes that justifies symbolism and confirms the primacy of intellectual intuition over discursive thought.

The keystone of all Christian cosmology and the element that renders possible the linking of the Biblical myth with the Greek heritage is the evangelical doctrine of the Logos as source of both existence and knowledge. This doctrine, which in itself transcends the plane of cosmology—the Gospels contain hardly any cosmological elements—constitutes nonetheless its spiritual axis; it is through this doctrine that the science of the created is connected with the knowledge of the uncreated. It is thus through its link with metaphysics—comprised in this case in the Johannine doctrine of the Word—that cosmology is in agreement with theology. It is first of all a prolongation of gnosis; thereafter an *ancilla theologiae*.

The same can be said of all traditional cosmologies and in particular of those belonging to Islam and Judaism; their immutable axis is always a revealed doctrine of the Spirit or Intellect, whether this be conceived as uncreated (as in the case of the Word) or as created (as with the first Intellect) or as having two aspects, one created and the other uncreated.[6]

We know that there were frequent exchanges between the Christian, Moslem, and Jewish cosmologists, and the same certainly occurred between the Hellenistic cosmologists and certain Asiatic civilizations; but it goes without saying (as René Guénon pointed out) that the family resemblance between all the traditional cosmologies had generally speaking nothing to do with historical borrowings, for in the first place there is the nature of things and, after that, there is intuitive knowledge. This knowledge, as we have said, must be vivified by a sacred science, the written and oral repository of a divine revelation. Be that as it may, everything is definitively contained within our own soul, whose lower ramifications are identified with the domain of the senses, but whose root reaches to pure Being and the supreme Essence, so that man grasps within himself the axis of the cosmos. He can "measure" the whole of its "vertical" dimension, and in this connection his knowledge of the world can be adequate, in spite of the fact that he will always be ignorant of

6 Ibn 'Arabî says the same in speaking of *ar-Rûh*, the Universal Spirit, in accordance with certain Koranic formulations. As for the first Intellect *(nous)* of Plotinus, it can also be regarded under these two aspects; the Plotinian doctrine of divine emanations does not introduce the distinction created-uncreated.

much, or even nearly all, of its "horizontal" extension. It is thus perfectly possible for traditional cosmology to possess, as it does, a knowledge that is real—and incomparably more vast and profound than that offered by the modern empirical sciences—while retaining childlike (or, more precisely, "human") opinions about realities of the physical order.

Western cosmology fell out of favor the moment the ancient geocentric system of the world was replaced by the heliocentric system of Copernicus. For that to be possible, cosmology had to be reduced to mere cosmography; thus the form was confused with the content, and the one was rejected with the other. In reality, the medieval conception of the physical world, of its ordonnance and of its extension, not only corresponded to a natural, and therefore realistic vision of things, it also expressed a spiritual order in which man had his organic place.

Let us pause for a moment at this vision of the world, known to us especially through the poetic works of Dante.[7] The planetary heavens and the heaven of fixed stars that surrounds them were presented as so many concentric spheres—"the vaster they are, the greater their virtue," as Dante explained—whose extreme limit, the invisible heaven of the Empyrean, is identified both with universal space and pure duration. Spatially, it represents a sphere of unlimited radius, and temporally, it is the background of all movement. Its continual rotation bears along with it all inferior movements, which are measured in relation to it, though it cannot itself be measured in any absolute way, since time cannot be divided except by reference to the marking out of a movement in space.

These spheres symbolize the higher states of consciousness and, more exactly, the modalities of the soul which, while still contained within the integral individuality, are more and more irradiated by the Divine Spirit. It is the Empyrean, the "threshold" between time and non-time, that represents the extreme limit of the individual or formal world. It is in crossing this limit that Dante obtains a new vision, one that is to some extent inverse to the cosmic order. Up to

7 There has been much discussion as to whether the *Divine Comedy* was influenced by an Islamic model; though possible in itself, it is not necessarily so, given that the symbolism in question resulted on the one hand from the spiritual realities themselves and on the other from the Ptolemaic system that was common to both Christian and Moslem civilizations in the Middle Ages.

that point the hierarchy of existence, which goes from corporeal to spiritual, expresses itself through a gradual expansion of space, the container being the cause and master of the contained. At this point the Divine Being reveals itself as the center around which the angels revolve in closer and closer choirs. In reality there is no symmetry between the two orders, planetary and angelic, for God is at one and the same time the center and container of all things. It is the physical order alone, that of the starry firmament, that represents the reflection of the superior order.

As for the circles of hell, which Dante[8] describes as a pit sunk into the earth as far as the "point toward which all heaviness tends," they are not the inverse reflection but the opposite of the heavenly spheres. They are, as it were, these spheres overturned, whereas the mountain of purgatory, which the poet tells us was formed from the earth cast up by Lucifer in the course of his fall towards the center of gravity, is properly speaking a compensation for hell. By this localization of hell and purgatory, Dante did not intend to establish a geography; he was not deluded concerning the provisional character of the symbolism, although he obviously believed in the geocentric system of Ptolemy.

The heliocentric system itself admits of an obvious symbolism, since it identifies the center of the world with the source of light. Its rediscovery by Copernicus,[9] however, produced no new spiritual vision of the world; rather it was comparable to the popularization of an esoteric truth. The heliocentric system had no common measure with the subjective experiences of people; in it man had no organic place. Instead of helping the human mind to go beyond itself and to consider things in terms of the immensity of the cosmos, it only encouraged a materialistic Prometheanism which, far from being superhuman, ended by becoming inhuman. Strictly speaking, a modern cosmology does not exist, in spite of the misuse of language whereby the modern science of the sensible universe is

8 With regard to the symbolical localization of the hells, medieval authors differ and seem to contradict one another. For Dante, the hells are situated beneath the earth, which means that they correspond to inferior states; for others, and especially for certain Moslem cosmologists, they are to be found "between heaven and earth," in other words, in the subtle world.

9 For it is not a case of an unprecedented discovery. Copernicus himself refers to Nicetas of Syracuse as also to certain quotations in Plutarch.

called cosmology. In fact, the modern science of nature expressly limits itself to the corporeal domain alone, which it isolates from the total cosmos while considering things in their purely spatial and temporal phenomenality, as if supra-sensible reality with its differing levels were nothing at all and as if that reality were not knowable by means of the intellect, in which it is analogically inherent by virtue of the correspondence between the macrocosm and the microcosm. But the point we wish to stress here is the following: scientism is an objectivism which purports to be mathematical and exclusive. Because of this, it behaves as if the human subject did not exist, or as if this subject were not the subtle mirror indispensable for the phenomenal appearance of the world. It is deliberately ignored that the subject is the guarantor of the logical continuity of the world and, in its intellectual essence, the witness of all objective reality.

In fact, a knowledge that is "objective," and thus independent of particular subjectivities presupposes immutable criteria, and these could not exist if there were not in the individual subject itself an impartial background, a witness transcending the individual, in other words the intellect. After all, knowledge of the world presupposes the underlying unity of the knowing subject, so that one might say of a voluntarily agnostic science what Meister Eckhart said about atheists: "The more they blaspheme God, the more they praise Him." The more science affirms an exclusively "objective" order of things, the more it manifests the underlying unity of the intellect or spirit; it does this indirectly, unconsciously, and in spite of itself—in other words, contrarily to its own thesis—but when all is said and done, it proclaims in its own way what it purports to deny. In the perspective of scientism, the total human subject— composed of sensibility, reason, and intellect—is illusorily replaced by mathematical thought alone. According to a scientist of the present century,[10] "All true progress in natural science consists in its disengaging itself more and more from subjectivity and in bringing out more and more clearly what exists independently of human conception, without troubling itself with the fact that the result has no longer anything but the most distant resemblance to what the original perception took for real." According to this declaration,

---

10 Sir James Jeans, *The New Background of Science* (Cambridge, 1933).

which is considered to be authoritative, the subjectivity from which one is to break loose is not reducible to the intrusion of sensorial accidents and emotional impulsions into the order of objective knowledge; it is the complete "human conception" of things—in other words, both direct sensory perception and its spontaneous assimilation by the imagination—which is called in question; only mathematical thought is allowed to be objective or true. Mathematical thought in fact allows a maximum of generalization while remaining bound to number, so that it can be verified on the quantitative plane; but it in nowise includes the whole of reality as it is communicated to us by our senses. It makes a selection from out of this total reality, and the scientific prejudice of which we have just been speaking regards as unreal everything that this selection leaves out. Thus it is that those sensible qualities called "secondary," such as colors, odors, savors, and the sensations of hot and cold, are considered to be subjective impressions implying no objective quality, and possessing no other reality than that belonging to their indirect physical causes, as for example, in the case of colors, the various frequencies of light waves: "Once it be admitted that in principle the sensible qualities cannot automatically be looked on as being qualities of the things themselves, physics offers us an entirely homogeneous and certain system, which answers every question as to what really underlies those colors, sounds, temperatures, etc."[11] What is this homogeneity but the result of a reduction of the qualitative aspects of nature to quantitative modalities? Modern science thus asks us to sacrifice a goodly part of what constitutes for us the reality of the world, and offers us in exchange mathematical formulae whose only advantage is to help us to manipulate matter on its own plane, which is that of quantity.

This mathematical selection from out of total reality does not only eliminate the "secondary" qualities of perception, it also removes what the Greek philosophers and the Scholastics called "form," in other words, the qualitative "seal" imprinted on matter by the unique essence of a being or a thing. For modern science, the essential form does not exist: "Some rare Aristotelians," writes a theoretician of modern science,[12] "still perhaps think they can intu-

---

11 B. Bavink, *Hauptfragen der heutigen Naturphilosophie* (Berlin, 1928).
12 Josef Geiser, *Allgemeine Philosophie des Seins und der Natur* (Munster 1. W., 1915).

itively attain, through some illumination by the active intellect, the essential ideas of the things of nature; but this is nothing but a beautiful dream. . . . The essences of things cannot be contemplated, they must be discovered by experience, by means of a laborious work of investigation." To this a Plotinus, an Avicenna, or a St. Albert the Great would reply that in nature there is nothing more evident than the essences of things, since these manifest themselves in the "forms" themselves. Only, they cannot be discovered by a "laborious work of investigation" nor measured quantitatively; in fact the intuition that grasps them relies directly on sensory perception and imagination, inasmuch as the latter synthesizes the impressions received from outside.

In any case, what is this human reason that tries to grasp the essences of things by a "laborious work of investigation"? Either this faculty of reason is truly capable of attaining its objects, or it is not. We know that reason is limited, but we also know that it is able to conceive truths that are independent of individuals, and that therefore a universal law is manifested in it. If human intelligence is not merely "organized matter"—in which case it would not be intelligence—this means that it necessarily participates in a transcendent principle. Without entering into a philosophical discussion on the nature of reason, we can compare the relationship between it and its supra-individual source (which medieval cosmology calls the "active intellect" and, in a more general sense, the "first intellect") to the relationship between a reflection and its luminous source, and this image will be both more ample and more accurate than any philosophical definition. A reflection is always limited by the nature of its plane of reflection—in the case of reason, this plane is the mind and, in a more general sense, the human psyche—but the nature of light remains essentially the same, in its source as in its reflection. The same applies to the spirit, whatever be the formal limits a particular plane of reflection imposes on it. Now spirit is essentially and wholly knowledge; in itself it is subject to no external constraint, and in principle nothing can prevent it from knowing itself and at the same time knowing all the possibilities contained within itself. Therein lies the mode of access, not to the material structure of things in particular and in detail, but to their permanent essences.

All true cosmological knowledge is founded on the qualitative aspect of things, in other words, on "forms" inasmuch as these are the mark of the essence. Because of this, cosmology is both direct and speculative, for it grasps the qualities of things in a direct way, and does not call them in question, and at the same time it disengages these qualities from their particular attachments so as to be able to consider them at their different levels of manifestation. In this way, the universe reveals its internal unity and at the same time shows the inexhaustible spectrum of its aspects and dimensions. That this vision should often have something poetic about it is obviously not to its detriment, since all genuine poetry comprises a presentiment of the essential harmony of the world; it was in this sense that Mohammed could say: "Surely there is a part of wisdom in poetry."

If one can reproach this vision of the world for being more contemplative than practical and for neglecting the material connections of things (which in reality is hardly a reproach), it can on the other hand be said about scientism that it empties the world of its qualitative sap. The traditional vision of things is above all "static" and "vertical." It is static because it refers to constant and universal qualities, and it is vertical in the sense that it attaches the lower to the higher, the ephemeral to the imperishable. The modern vision, on the contrary, is fundamentally "dynamic" and "horizontal"; it is not the symbolism of things that interests it, but their material and historical connections.

The great argument in favor of the modern science of nature—an argument that counts for much in the eyes of the crowd (whatever may be the reservations of men of science themselves)—is its technical application; this, it is believed, proves the validity of the scientific principles,[13] as if a fragmentary and in some respects problematical efficacy could be a proof of their intrinsic and total value. In reality, modern science displays a certain number of fissures that are due to the fact that the world of phenomena is indefinite and that therefore no science can ever hope to exhaust it; these fissures derive above all from modern science's systematic exclusion of all the non-corporeal dimensions of reality. They manifest themselves right down to the foundations of modern science, and in domains as seemingly "exact" as that of physics; they become gaping cracks

---

13 It is a fact, however, that most of the great technical inventions were effected on the basis of inadequate and even false theories.

when one turns to the disciplines connected with the study of life, not to mention psychology, where an empiricism that is relatively valid in the physical order encroaches in bizarre fashion on a foreign field. These fissures, which do not merely affect the theoretical domain, are far from harmless; on the contrary, in their technical consequences, they constitute so many seeds of catastrophe.

Because the mathematical conception of things inevitably participates in the schematic and discontinuous character of number, it neglects, in the vast web of nature, everything that consists of pure continuity and of relations subtly kept in balance. Now, continuity and equilibrium exist before discontinuity and before crisis; they are more real than these latter, and incomparably more precious.

(II) MODERN PHYSICS

In modern physics the space in which the heavenly bodies move, as also the space traversed by the trajectories of the minutest bodies such as electrons, is conceived as a void. The purely mathematical definition of the spatial and temporal relationships between various bodies great or small is thereby rendered easier. In reality, a corporeal "point" "suspended" in a total void would have no relationship whatever with any other corporeal "points"; it would, so to speak, fall back into nothingness. One blithely speaks of "fields of force," but by what are these fields supported? A totally empty space cannot exist; it is only an abstraction, an arbitrary idea that serves only to show where mathematical thinking can lead when arbitrarily detached from a concrete intuition of things.

According to traditional cosmology, ether fills all space without distinction. We know that modern physics denies the existence of ether, since it has been established that it offers no resistance to the rotatory movement of the earth; but it is forgotten that this quintessential element which is at the basis of all material differentiations, is not itself distinguished by any particular quality, so that it offers no opposition to anything whatsoever. It represents the continuous ground whence all material discontinuities detach themselves.

If modern science accepted the existence of ether, it might perhaps find an answer to the question whether light is propagated as a wave or as a corpuscular emanation; most probably its movement is neither one nor the other, and its apparently contradictory properties are explainable by the fact that it is most directly attached to ether and participates in the indistinctly continuous nature of the latter.

An indistinct continuum cannot be divided into a series of like units; if it does not necessarily escape from time or space, it nevertheless eludes graduated measurements. This is especially true of the speed of light, which always appears the same, independently of the movement of its observer, whether the latter moves in the same direction or in the opposite direction. The speed of light thus represents a limit value; it can neither be overtaken nor caught up with by any other movement, and this is like the physical expression of the simultaneity proper to the act of the intelligible light.

We know that the discovery of the fact that the speed of light, when measured both in the direction of the rotation of the earth and in the direction opposite to that rotation, is invariable, has confronted modern astronomers with the alternative either of accepting the immobility of the earth or else of rejecting the usual notions of time and space. Thus it was that Einstein was led into considering space and time as two relative dimensions, variable in function of the state of movement of the observer, the only constant dimension being the speed of light. The latter would everywhere and always be the same, whereas time and space vary in relation to one another: it is as if space could shrink in favor of time, and inversely.

If it be admitted that a movement is definable in terms of a certain relationship of time and space, it is contradictory to maintain that it is a movement, that of light, that measures space and time. It is true that on a quite different plane—when it is a question of the intelligible light—the image of light "measuring" the cosmos and realizing it thereby is not devoid of deep meaning. But what we have in view here is the physical order, which alone is considered, and with good cause, by Einstein's theory; it is therefore in this context that we will put the following question: what is this famous "constant number" that is supposed to express the speed of light? How can movement having a definite speed—and its definition will always be a relationship between space and time—itself be a quasi-"absolute" measure of these two conditions of the physical world? Is there not here a confusion between the principial and quantitative domains? That the movement of light is the fundamental "measure" of the corporeal world we willingly believe, but why should this measure itself be a number, and even a definite number? Moreover, do the experiments which are supposed to prove the constant character of the speed of light really get beyond the earthly sphere, and do they not imply both space and time as usually imagined by us? Thus "300,000 km per second" is stated to be the speed of light, and it is held that here is a value which, if it be not necessarily everywhere expressed in this manner, does nonetheless remain constant throughout the physical universe. The astronomer who counts, by referring to the lines of the spectrum, the light-years separating us from the nebula of Andromeda, supposes without more ado that the universe is everywhere "woven" in the same manner. Now, what would happen if the constant character of the speed of light ever came to be doubted—and there is every likelihood that it will be sooner or later—so that the only fixed pivot of Einstein's theory would fall down? The whole modern conception of the universe would immediately dissolve like a mirage.

We are told that reality does not necessarily correspond with our inborn conceptions of time and space; but at the same time it is never doubted for a moment that the physical universe conforms with certain mathematical formulas which necessarily proceed from axioms that are no less inborn.

In the same order of ideas mention must also be made of the theory according to which interstellar space is not the space of Euclid, but a space that does not admit the Euclidean axiom regarding parallel lines. Such a space, it is said, flows back on itself, without its being possible to assign to it a definite curve. One might see in this theory an expression of spatial indefinitude, since, in fact space is neither finite nor infinite, something which the Ancients indicated by comparing space to a sphere whose radius exceeds every measure, and which itself is contained in the Universal Spirit. But this is not how modern theoreticians understand things, for they declare that our immediate conception of space is quite simply false and incomplete, and that we must therefore familiarize ourselves with non-Euclidean space, which, they say, is accessible to a disciplined imagination. Now this is simply not true, for non-Euclidean space is accessible only indirectly, namely, from the starting-point of Euclidean space, which thus remains the qualitative model for every conceivable kind of space. In this case, as in many others, modern science tries mathematically to go beyond the logic inherent in the imagination, and then to violate this by dint of mathematical principles, as if every intellectual faculty other than purely mathematical thought were suspect.

In conformity with this mathematical schematism, matter itself is conceived as being discontinuous, for atoms, and their constituent particles, are supposed to be even more isolated in space than are the stars. Whatever the current conception of the atomic order may be—and theories on this subject change at a disconcerting speed—it is always a case of groupings of corporeal "points."

Let us here recall the traditional doctrine of matter:[14] it is from the starting-point of "first matter" that the world is constituted, by successive differentiation, under the "non-acting" action of the form-bestowing Essence; but this *materia prima* is not tangible matter, it underlies all finite existence, and even its nearest modality, *materia signata quantitate,* which is the basis of the corporeal world, is not manifested as such. According to a most judicious expression of Boethius,[15] it is by its "form"—in other words, its qualitative aspect—that a thing is known, "form being like a light by means of which we know what a thing is," Now *materia* as such is precisely that which is not yet formed and which by that very fact eludes all distinctive knowing. The world that is accessible to distinctive knowledge thus extends between two poles that are unmanifested as such (the form-bestowing Essence and undifferentiated *materia)* just as the range of colors in the spectrum unfolds through the refraction of white—and therefore colorless—light in a medium that is also colorless.

Modern science, which despite its pragmatism is not behind-hand in claiming to offer a complete and comprehensive explanation of the sensible universe, strives to reduce the whole qualitative richness of this universe to a certain structure of matter, conceived as a variable grouping of minute bodies, whether these be defined as genuine bodies or as simple "points" of energy. This means that all the "bundles" of sensible qualities, everything that constitutes the world for us, except space and time, have to be reduced, scientifically speaking, to a series of atomic "models" definable in terms

---

14 "René Guénon, *The Reign of Quantity and the Signs of the Times* (Sophia Perennis, Ghent, New York, 2001), Chapter 2, *"Materia Signata Quantitate."*
15 *De Unitate et Uno.*

of the number, mass, trajectories, and speeds of the minute bodies concerned. It is obvious that this reduction is in vain, for although these "models" still comprise certain qualitative elements—if only their imaginary spatial form—it is nonetheless a question of the reduction of quality to quantity—and quantity can never comprehend quality.

On the other hand, the elimination of the qualitative aspects in favor of a tighter and tighter mathematical definition of atomic structure must necessarily reach a limit, beyond which precision gives way to the indeterminate. This is exactly what is happening with modern atomist science, in which mathematical reflection is being more and more replaced by statistics and calculations of probability, and in which the very laws of causality seem to be facing bankruptcy. If the "forms" of things are "lights," as Boethius said, the reduction of the qualitative to the quantitative can be compared to the action of a man who puts out all the lights the better to scrutinize the nature of darkness.

Modern science can never reach that matter that is at the basis of this world. But between the qualitatively differentiated world and undifferentiated matter lies something like an intermediate zone—and this is chaos. The sinister dangers of atomic fission are but one signpost indicating the frontier of chaos and dissolution.

(III) TRADITIONAL SYMBOLISM & MODERN EMPIRICISM

If the ancient cosmogonies seem childish when one takes their symbolism literally—and this means not understanding them—modern theories about the origin of the world are frankly absurd. They are so, not so much in their mathematical formulations, but because of the total unawareness with which their authors set themselves up as sovereign witnesses of cosmic becoming, while at the same time claiming that the human mind itself is a product of this becoming. What connection is there between that primordial nebula—that vortex of matter whence they wish to derive earth, life, and man—and this little mental mirror that loses itself in conjectures (since for the scientists intelligence amounts to no more than this) and yet feels so sure of discovering the logic of things within itself? How can the effect make judgements regarding its own cause? And if there exist constant laws of nature—those of causality, number, space, and time—and something which, within ourselves, has the right to say "this is true and this is false," where is the guarantee of truth, either in the object or in the subject? Is the nature of our mind merely a little foam on the waves of the cosmic ocean, or is there to be found deep within it a timeless witness of reality?

Some protagonists of the theories in question will perhaps say that they are concerned only with the physical and objective domain, without seeking to prejudge the domain of the subjective. They can perhaps cite Descartes, who defined spirit and matter as two realities, coordinated by Providence, but separated in fact. In point of fact, this division of reality into watertight compartments served to prepare people's minds to leave aside everything that is not of the physical order, as if man were not himself proof of the complexity of the real.

The man of antiquity, who pictured the earth as an island surrounded by the primordial ocean and covered by the dome of heaven, and the medieval man, who saw the heavens as concentric spheres extending from the earth (viewed as the center) to the limitless sphere of the Divine Spirit, were no doubt mistaken regarding the true disposition and proportions of the sensible universe. On the other hand, they were fully conscious of the fact—infinitely more important—that this corporeal world is not the whole of reality, and that it is as if surrounded and pervaded by a reality, both greater and more subtle, that in its turn is contained in the Spirit; and they knew, indirectly or directly, that the world in all its extension disappears in the face of the Infinite.

Modern man knows that the earth is only a ball suspended in a bottomless abyss and carried along in a dizzy and complex movement, and that this movement is governed by other celestial bodies incomparably larger than this earth and situated at immense distances from it. He knows that the earth on which he lives is but a grain in comparison with the sun, which itself is but a grain amidst other incandescent stars, and that all is in motion. An irregularity in this assemblage of sidereal movements, an interference from a star foreign to our planetary system, a deviation of the sun's trajectory, or any other cosmic accident, would suffice to make the earth unsteady in its rotation, to trouble the course of the seasons, to change the atmosphere, and to destroy mankind. Modern man also knows that the smallest atom contains forces which, if unleashed, could involve the earth in an almost instantaneous conflagration. All of this, from the "infinitely small" to the "infinitely great," presents itself, from the point of view of modern science, as a mechanism of unimaginable complexity, the functioning of which is only due to blind forces.

In spite of this, the man of our time lives and acts as if the normal and habitual operation of the rhythms of nature were something that was guaranteed to him. In actual practice, he thinks neither of the abysses of the stellar world nor of the terrible forces latent in every particle of matter. He sees the sky above him like any child sees it, with its sun and its stars, but the remembrance of the astronomical theories prevents him from recognizing divine signs in them. The sky for him is no longer the natural expression of the Spirit that enfolds and illuminates the world. Scientific knowledge has substituted itself for this "naïve" and yet profound vision, not as a new consciousness of a vaster cosmic order, an order of which man forms part, but as an estrangement, as an irremediable disarray before abysses that no longer have any common measure with him. For nothing now reminds him that in reality this whole universe is contained within himself, not of course in his individual being, but in the spirit or intellect that is within him and that is both greater than himself and the whole phenomenal universe.

(IV) EVOLUTIONISM

The least phenomenon participates in several continuities or cosmic dimensions incommensurable in relation to each other; thus, ice is water as regards its substance—and in this respect it is indistinguishable from liquid water or water vapor—but as regards its state it belongs to the class of solid bodies. Similarly, when a thing is constituted by diverse elements, it participates in their natures while being different from them. Cinnabar, for example, is a synthesis of sulphur and mercury; it is thus in one sense the sum of these two elements, but at the same time it possesses qualities that are not to be found in either of these two substances. Quantities can be added to one another, but a quality is never merely the sum of other qualities. By mixing the colors blue and yellow, green is obtained; this third color is thus a synthesis of the other two, but it is not the product of a simple addition, for it represents at the same time a chromatic quality that is new and unique in itself.

There is here something like a "discontinuous continuity," which is even more marked in the biological order, where the qualitative unity of an organism is plainly distinguishable from its material composition. The bird that is born from the egg is made from the same elements as the egg, but it is not the egg. Likewise, the butterfly that emerges from a chrysalis is neither that chrysalis nor the caterpillar that produced it. A kinship exists between these various organisms, a genetic continuity, but they also display a qualitative discontinuity, since between the caterpillar and the butterfly there is something like a rupture of level.

At every point in the cosmic web there is thus a warp and a woof that intersect one another, and this is indicated by the traditional symbolism of weaving, according to which the threads of the warp, which hang vertically on the primitive loom, represent the permanent essences of things—and thus also the essential qualities and forms—while the woof, which binds horizontally the threads of the warp, and at the same time covers them with its alternating waves, corresponds to the substantial or "material" continuity of the world.[16]

The same law is expressed by classical hylomorphism, which distinguishes the "form" of a thing or being—the seal of its essential unity—from its "matter," namely the plastic substance which receives this seal and furnishes it with a concrete and limited existence. No modern theory has ever been able to replace this ancient theory, for the fact of reducing the whole plenitude of the real to one or other of its "dimensions" hardly amounts to an explanation of it. Modern science is ignorant above all of what the Ancients designated by the term "form," precisely because it is here a question of a non-quantitative aspect of things, and this ignorance is not unconnected with the fact that modern science sees no criterion in the beauty or ugliness of a phenomenon: the beauty of a thing is the sign of its internal unity, its conformity with an indivisible essence, and thus with a reality that will not let itself be counted or measured.

It is necessary to point out here that the notion of "form" necessarily includes a twofold meaning: on the one hand it means the delimitation of a thing, and this is its most usual meaning; in this connection, form is situated on the side of matter or, in a more

---

16 René Guénon, *The Symbolism of the Cross* (Sophia Perennis, Ghent, New York, 2001) chapter on the symbolism of weaving.

general sense, on the side of plastic substance, which limits and separates realities.[17] On the other hand, "form" understood in the sense given to it by the Greek philosophers and following them the Scholastics, is the aggregate of qualities pertaining to a being or a thing, and thus the expression or the trace of its immutable essence.

The individual world is the "formal" world because it is the domain of those realities that are constituted by the conjunction of a "form" and a "matter," whether subtle or corporeal. It is only in connection with a "matter," a plastic substance, that "form" plays the role of a principle of individuation; in itself, in its ontological basis, it is not an individual reality but an archetype, and as such beyond limitations and change. Thus a species is an archetype, and if it is only manifested by the individuals that belong to it, it is nevertheless just as real, and even incomparably more real, than they. As for the rationalist objection that tries to prove the absurdity of the doctrine of archetypes by arguing that a multiplication of mental notions would imply a corresponding multiplication of archetypes—leading to the idea of the idea of the idea, and so on—it quite misses the point, since multiplicity can in nowise be transposed onto the level of the archetypal roots. The latter are distinguished in a principial way, within Being and by virtue of Being; in this connection, Being can be envisaged as a unique and homogeneous crystal potentially containing all possible crystalline forms.[18] Multiplicity and quantity thus only exist at the level of the "material" reflections of the archetypes.

From what has just been said, it follows that a species is in itself an immutable "form"; it cannot evolve and be transformed into another species, although it may include variants, which are diverse "projections" of a unique essential form, from which they can never be detached, any more than the branches of a tree can be detached from the trunk.

17 In Hindu parlance, the distinction *nâma-rupa*, "name and form," is related to this aspect of the notion under study, "name" here standing for the essence of a being or thing, and "form" for its limited and outward existence.

18 It is self-evident that all the images that one can offer of the non-separative distinction of the possibilities contained in Being must remain imperfect and paradoxical.

It has been justly said[19] that the whole thesis of the evolution of species, inaugurated by Darwin, is founded on a confusion between species and simple variation. Its advocates put forward as the "bud" or the beginning of a new species what in reality is no more than a variant within the framework of a determinate specific type. This false assimilation is, however, not enough to fill the numberless gaps that occur in the paleontological succession of species; not only are related species separated by profound gaps, but there do not even exist any forms that would indicate any possible connection between different orders such as fish, reptiles, birds, and mammals. One can doubtless find some fishes that use their fins to crawl onto a bank, but one will seek in vain in these fins for the slightest beginning of that articulation which would render possible the formation of an arm or a paw. Likewise, if there are certain resemblances between reptiles and birds, their respective skeletons are nonetheless of a fundamentally different structure. Thus, for example, the very complex articulation in the jaws of a bird, and the related organization of its hearing apparatus, pertain to an entirely different plan from the one found in reptiles; it is difficult to conceive how one might have developed from the other.[20] As for the famous fossil bird *Archaeopteryx,* it is fairly and squarely a bird, despite the claws at the end of its wings, its teeth, and its long tail.[21]

In order to explain the absence of intermediate forms, the partisans of transformism have sometimes argued that these forms must have disappeared because of their very imperfection and precariousness; but this argument is plainly in contradiction with the principle of selection that is supposed to be the operative factor in the evolution of species: the trial forms should be incomparably more numerous than the ancestors having already acquired a definitive form. Besides, if the evolution of species represents, as is declared, a gradual and continual process, all the real links in the chain—therefore all those that are destined to be followed—will be

---

19 Douglas Dewar, *The Transformist Illusion* (Dehoff Publications, Murfreesboro, Tennessee, 1957 [*Editor's note*: more recently reprinted by Sophia Perennis et Universalis, 1995.]). See also Louis Bounoure, *Déterminisme et Finalité* (Collection Philosophie, Flammarion, Paris).

20 Dewar, *The Transformist Illusion.*

21 *Ibid.*

both endpoints and intermediaries, in which case it is difficult to see why the ones would be much more precarious than the others.[22]

The more conscientious among modern biologists either reject the transformist theory, or else maintain it as a "working hypothesis," being unable to conceive any genesis of species that would not be situated on the "horizontal line" of a purely physical and temporal becoming. For Jean Rostand,

> the world postulated by transformism is a fairy-like world, phantasmagoric, surrealistic. The chief point, to which one always returns, is that we have never been present, even in a small way, at *one* authentic phenomenon of evolution . . . we keep the impression that nature today has nothing to offer that might be capable of reducing our embarrassment before the veritably organic metamorphoses implied in the transformist thesis. We keep the impression that, in the matter of the genesis of species as in that of the genesis of life, the forces that constructed nature are now absent from nature . . .[23]

Even so, this biologist sticks to the transformist theory:

> I firmly believe—because I see no means of doing otherwise—that mammals have come from lizards, and lizards from fish; but when I declare and when I think such a thing, I try not to avoid seeing its indigestible enormity and I prefer to leave vague the origin of these scandalous metamorphoses rather than add to their improbability that of a ludicrous interpretation.[24]

All that paleontology proves to us is that the various animal forms, such as are shown by fossils preserved in successive earthly

---

22 Teilhard de Chardin *(The Human Phenomenon,* p. 129) writes on this subject: "Nothing is by nature so delicate and fugitive as a beginning. As long as a zoological group is young, its characteristics remain undecided. Its dimensions are weak. Relatively few individuals compose it, and these are rapidly changing. Both in space and duration, the peduncle (or the bud, which comes to the same thing) of a living branch corresponds to a minimum of differentiation, expansion, and resistance. How then is time going to act on this weak zone? Inevitably by destroying it in its vestiges." This reasoning, which abusively exploits the purely external and conventional analogy between a genealogical "tree" and a real plant, is an example of the "imaginative abstraction" that characterizes this author's thought.

23 *Le Figaro Littéraire,* April 20, 1957.

24 *Ibid.*

layers, made their appearance in a vaguely ascending order, going from relatively undifferentiated organisms—but not simple ones[25]—to ever more complex forms, without this ascension representing, however, an unequivocal and continuous line. It seems to move in jumps; in other words, whole categories of animals appear all at once, without real predecessors. What does this order mean? Simply that, on the material plane, the simple or relatively undifferentiated always precedes the complex and differentiated. All "matter" is like a mirror that reflects the activity of the essences, while also inverting it; this is why the seed comes before the tree and the bud before the flower, whereas in the principial order the perfect "forms" pre-exist. The successive appearance of animal forms according to an ascending hierarchy therefore in nowise proves their continual and cumulative genesis.[26]

On the contrary, what links the various animal forms to one another is something like a common model, which reveals itself more or less through their structures and which is more apparent in the case of animals endowed with superior consciousness such as birds and mammals. This model is expressed especially in the symmetrical disposition of the body, in the number of extremities and sensory organs, and also in the general form of the chief internal organs. It might be suggested that the design and number of certain organs, and especially those of sensation, simply correspond to the terrestrial surroundings; but this argument is reversible, because those surroundings are precisely what the sensory organs grasp and delimit. In fact, the model underlying all animal forms establishes the analogy between the microcosm and the macrocosm. Against

25 The electron microscope has revealed the surprising complexity of the functions at work within a unicellular being.

26 The most commonly mentioned example in favor of the transformist thesis is the hypothetical genealogy of the *Equidae*. Charles Depéret criticizes it as follows: "Geological observation establishes in a formal manner that no gradual passage took place between these genera; the last *Palaeotherium* had for long been extinct, without having transformed itself, when the first *Architherium* made its appearance, and the latter disappeared in its turn, without modification, before being suddenly replaced by the invasion of the *Hipparion*." (*Les Transformations du Monde animal*, p. 107) To this it can be added that the supposed primitive forms of the horse are hardly to be observed in equine embryology, though the development of the embryo is commonly looked on as a recapitulation of the genesis of the species.

the background of this common cosmic pattern the differences between species and the gaps that separate them are all the more marked.

Instead of "missing links," which the partisans of transformism seek in vain, nature offers us, as if in irony, a large variety of animal forms which, without transgressing the pre-established framework of a species, imitate the appearance and habits of a species or order foreign to them. Thus, for example, whales are mammals, but they assume the appearance and behavior of fishes; hummingbirds have the appearance, iridescent colors, flight, and mode of feeding of butterflies; the armadillo is covered with scales like a reptile, although it is a mammal; and so on. Most of these animals with imitative forms are higher species that have taken on the forms of relatively lower species, a fact which a priori excludes an interpretation of them as intermediary links in an evolution. As for their interpretation as forms of adaptation to a given set of surroundings, this seems more than dubious, for what could be, for example, the intermediate forms between some land mammal or other and the dolphin?[27] Among these "imitative" forms, which constitute so many extreme cases, we must also include the fossil bird *Archaeopteryx* mentioned above.

Since each animal order represents an archetype that includes the archetypes of the corresponding species, one might well ask oneself whether the existence of "imitative" animal forms does not contradict the immutability of the essential forms; but this is not the case, for the existence of these forms demonstrates, on the contrary, that very immutability by a logical exhausting of all the possibilities inherent in a given type or essential form. It is as if nature, after bringing forth fishes, reptiles, birds, and mammals, with their distinctive characteristics, wished still to show that she was able to produce an animal like the dolphin which, while being a true mammal, at the same time possesses almost all the faculties of a fish, or a creature like the tortoise, which possesses a skeleton covered by

---

27 On the subject of the hypothetical transformation of a land animal into the whale, Douglas Dewar writes: "I have often challenged transformists to describe plausible ancestors situated in the intermediate phases of this supposed transformation" (*What the Animal Fossils Tell Us,* Trans. Vict. Instit, vol. LXXIV).

flesh, yet at the same time is enclosed in an exterior carapace after the fashion of certain mollusks.[28] Thus does nature manifest her protean power, her inexhaustible capacity for generation, while remaining faithful to the essential forms, which in fact are never blurred.

Each essential form—or each archetype—includes after its fashion all the others, but without confusion; it is like a mirror reflecting other mirrors, which reflect it in their turn.[29] In its deepest meaning the mutual reflection of types is an expression of the metaphysical homogeneity of Existence, or of the unity of Being.

Some biologists, when confronted with the discontinuity in the paleontological succession of species, postulate an evolution by leaps and, in order to make this theory plausible, refer to the sudden mutations observed in some living species. But these mutations never exceed the limits of an anomaly or a decadence, as for example the sudden appearance of albinos, or of dwarfs or giants; even when these characteristics become hereditary, they remain as anomalies and never constitute new specific forms.[30] For this to happen, it would be necessary for the vital substance of an existing species to serve as the "plastic material" for a newly manifested specific form; in practice, this means that one or several females of this existing species would suddenly bear offspring of a new species. Now, as the hermetist Richard the Englishman writes:

> Nothing can be produced from a thing that is not contained in it; for this reason, every species, every genus, and every natural order develops within the limits proper to it and bears fruits according to its own kind and not according to an essentially different order; everything that receives a seed must be of the same seed.[31]

Fundamentally, the evolutionist thesis is an attempt to replace, not simply the "miracle of creation," but the cosmogonic process—

---

28 It is significant that the tortoise, whose skeleton seems to indicate an extravagant adaptation to an animal "armored" state, appears all at once among the fossils, without evolution. Similarly, the spider appears simultaneously with its prey and with its faculty of weaving already developed.

29 This is the image used by the Sufi 'Abd al-Karîm al-Jîlî in his book *al-Insân al-Kâmil*, chapter on "Divine Unicity."

30 Bounoure, *Déterminisme et Finalité*.

31 Quoted in the *Golden Treatise*, Museum Hermeticum (Frankfurt, 1678).

largely suprasensory—of which the Biblical narrative is a Scriptural symbol; evolutionism, by absurdly making the greater derive from the lesser, is the opposite of this process, or this "emanation." (This term has nothing to do with the emanationist heresy, since the transcendence and immutability of the ontological principle are here in no wise called into question.) In a word, evolutionism results from an incapacity—peculiar to modern science—to conceive "dimensions" of reality other than purely physical ones; to understand the "vertical" genesis of species, it is worth recalling what René Guénon said about the progressive solidification of the corporeal state through the various terrestrial ages.[32] This solidification must obviously not be taken to imply that the stones of the earliest ages were soft, for this would be tantamount to saying that certain physical qualities—and in particular hardness and density—were then wanting; what has hardened and become fixed with time is the corporeal state taken as a whole, with the result that it no longer receives directly the imprint of subtle forms. Assuredly, it cannot become detached from the subtle state, which is its ontological root and which dominates it entirely, but the relationship between the two states of existence no longer has the creative character that it possessed at the origin; it is as when a fruit, having reached maturity, becomes surrounded by an ever harder husk and ceases to absorb the sap of the tree. In a cyclic phase in which corporeal existence had not yet reached this degree of solidification, a new specific form could manifest itself directly from the starting-point of its first "condensation" in the subtle or animic state;[33] this means that the different types of animals pre-existed at the level immediately superior to the corporeal world as non-spatial forms, but nevertheless clothed in a certain "matter," namely that of the subtle world. From there these forms "descended" into the corporeal state each time the latter was ready to receive them; this "descent" had the

32 René Guénon, *The Reign of Quantity and the Signs of the Times* (Sophia Perennis, Ghent, New York, 2001).

33 Concerning the creation of species in a subtle "proto-matter"—in which they still preserve an androgynous form, comparable to a sphere—and their subsequent exteriorization by "crystallization" in sensible matter (which is heavy, opaque, and mortal), see Frithjof Schuon, *Light on the Ancient Worlds* (World Wisdom Books, Bloomington, Indiana, 1984), Chapter 2, "In the Wake of the Fall," and *Form and Substance in the Religions* (World Wisdom, Bloomington, Indiana, 2002), Chapter 5, "The Five Divine Presences."

nature of a sudden coagulation and hence also the nature of a limitation and fragmentation of the original animic form. Indo-Tibetan cosmology describes this descent—which is also a fall—in the case of human beings under the form of the mythological combat of the *devas* and *asûras*: the *devas* having created man with a body that was fluid, protean, and diaphanous—in other words, in a subtle form—the *asûras* try to destroy it by a progressive petrification; it becomes opaque, gets fixed, and its skeleton, affected by the petrification, is immobilized. Thereupon the *devas,* turning evil into good, create joints, after having fractured the bones, and they also open the pathways of the senses, by piercing the skull, which threatens to imprison the seat of the mind. In this way the process of solidification stops before it reaches its extreme limit, and certain organs in man, such as the eye, still retain something of the nature of the non-corporeal states.[34]

In this story, the pictorial description of the subtle world must not be misunderstood. However, it is certain that the process of materialization, from the supra-sensory to the sensory, had to be reflected within the material or corporeal state itself, so that one can say without risk of error, that the first generations of a new species did not leave a mark in the great book of earthly layers; it is therefore vain to seek in sensible matter the ancestors of a species, and especially that of man.

Since the transformist theory is not founded on any real proof, its corollary and conclusion, namely the theory of the infra-human origin of man, remains suspended in the void. The facts adduced in favor of this thesis are restricted to a few groups of skeletons of disparate chronology: it happens that some skeletal types deemed to be more "evolved," such as "Steinheim man," precede others, of a seemingly more primitive character, such as "Neanderthal man," even though the latter was doubtless not so apelike as tendentious reconstructions would have us believe.[35]

If, instead of always putting the questions: at what point does humankind begin, and what is the degree of evolution of such and such a type regarded as being pre-human, we were to ask ourselves:

---

34 See Krasinsky, *Tibetische Medizin-Philosophie.*

35 In general, this domain of science has been almost smothered by tendentious theories, hoaxes, and imprudently popularized discoveries. See Dewar, *The Transformist Illusion.*

how far does the monkey go, things might well appear in a very different light, for a fragment from a skeleton, even one related to that of man, is hardly enough to establish the presence of that which constitutes man, namely reason, whereas it is possible to conceive of a great variety of anthropoid apes whose anatomies are more or less close to that of man.

However paradoxical this may seem, the anatomical resemblance between man and the anthropoid apes is explainable precisely by the difference—not gradual, but essential—that separates man from all other animals. Since the anthropoid form is able to exist without that "central" element that characterizes man—this "central" element manifesting itself anatomically by his vertical position, amongst other things—the anthropoid form must exist; in other words, there cannot but be found, at the purely animal level, a form that realizes in its own way—that is to say, according to the laws of its own level—the very plan of the human anatomy; the ape is a prefiguration of man, not in the sense of an evolutive phase, but by virtue of the law that decrees that at every level of existence analogous possibilities will be found.

A further question arises in the case of the fossils attributed to primitive men: did some of these skeletons belong to men we can look upon as being ancestors of men presently alive, or do they bear witness to a few groups that survived the cataclysm at the end of a terrestrial age, only to disappear in their turn before the beginning of our present humanity? Instead of primitive men, it might well be a case of degenerate men, who may or may not have existed alongside our real ancestors. We know that the folklore of most peoples speaks of giants or dwarfs who lived long ago, in remote countries; now, among these skeletons, several cases of gigantism are to be found.[36]

Finally, let it be recalled once more that the bodies of the most ancient men did not necessarily leave solid traces, either because their bodies were not yet at that point materialized or "solidified," or because the spiritual state of these men, along with the cosmic conditions of their time, rendered possible a resorption of the physical body into the subtle "body" at the moment of death.[37]

36 Like the Meganthrope of Java and the *Gigantopithecus* of China.
37 In some very exceptional cases—such as Enoch, Elijah, and the Virgin Mary—such a resorption took place even in the present terrestrial age.

We must now say a few words about a thesis, much in vogue today, which claims to be something like a spiritual integration of paleontology, but which in reality is nothing but a purely mental sublimation of the crudest materialism, with all the prejudices this includes, from belief in the indefinite progress of humanity to a leveling and totalitarian collectivism, without forgetting the cult of the machine that is at the center of all this; it will be apparent that we are here referring to Teilhardian evolutionism.[38] According to Teilhard de Chardin, who is not given to worrying over the gaps inherent in the evolutionist system and largely relies on the climate created by the premature popularization of the transformist thesis, man himself represents only an intermediate state in an evolution that starts with unicellular organisms and ends in a sort of global cosmic entity, united to God. The craze for trying to bring everything back to a single unequivocal and uninterrupted genetic line here exceeds the material plane and launches out wildly into an irresponsible and avid "mentalization" characterized by an abstraction clothed in artificial images which their author ends up by taking literally, as if he were dealing with concrete realities. We have already mentioned the imaginary genealogical tree of species, whose supposed unity is no more than a snare, being composed of the hypothetical conjunction of many disjointed elements. Teilhard amplifies this notion to his heart's content, in a manner that is purely graphic, by completing its branches—or "scales," as he likes

---

38 Teilhard's materialism is revealed in all its crudity, and all its perversity, when this philosopher advocates the use of surgical means to accelerate "collective cerebralization" (*Man's Place in Nature,* Harper & Row, New York, 1966). Let us also quote the further highly revealing words of the same author: "It is finally on the dazzling notion of Progress and on faith in Progress that today's divided humanity can be reformed. . . . Act one is over! We have access to the heart of the atom! Now come the next steps, such as the vitalization of matter by the building of super-molecules, the modeling of the human organism by hormones, the control of heredity and of the sexes by the play of genes and chromosomes, the readjustment and liberation by direct action of the springs laid bare by psychoanalysis, the awakening and taking hold of the still dormant intellectual and emotional forces in the human mass!" *(Planète III,* 1944), p. 30. Quite naturally, Teilhard proposes the fashioning of mankind by a universal scientific government—in short, all that is needed for the reign of the Antichrist. [*Editor's note:* The interested reader is referred to *Teilhardism and the New Religion: A Thorough Analysis of the Teachings of Pierre Teilhard de Chardin* (Rockford: Tan Books, 1988), by Wolfgang Smith, for a traditional critique of the views of the controversial Catholic priest and paleontologist.]

to call them—and by constructing a pinnacle in the direction of which humankind is supposed to be situated. By a similar sliding of thought from the abstract to the concrete, from the metaphorical to the supposedly real, he agglutinates, in one and the same pseudo-scientific outburst, the most diverse realities, such as mechanical laws, vital forces, psychic elements, and spiritual entities. Let us quote a characteristic passage:

> What explains the biological revolution caused by the appearance of Man, is an explosion of consciousness; and what, in its turn, explains this explosion of consciousness, is simply the passage of a privileged radius of "corpusculization," in other words, of a zoo-logical phylum, across the surface, hitherto impermeable, sepa-rating the zone of direct Psychism from that of reflective Psychism. Having reached, following this particular radius, a critical point of arrangement (or, as we say here, of enrolment), Life became hypercentered on itself, to the point of becoming capable of fore-sight and invention . . .[39]

Thus, "corpusculization" (which is a physical process) would have as its effect that a "zoological phylum" (which is no more than a figure) should pass across the surface (purely hypothetical) sepa-rating two psychic zones. . . . But we must not be surprised at the absence of *distinguos* in Teilhard's thinking since, according to his own theory, the mind is but a metamorphosis of matter!

Without stopping to discuss the strange theology of this author, for whom God himself evolves along with matter, and without daring to define what he thinks of the prophets and sages of antiq-uity and other "underdeveloped" beings of this kind, we will say the following: if man, in respect of both his physical nature and his spir-itual nature, were really nothing but a phase of an evolution going from the amoeba to the superman, how could he know objectively where he stands in all this? Let us suppose that this alleged evolu-tion forms a curve, or a spiral. The man who is but a fragment thereof—and let it not be forgotten that a "fragment" of a move-ment is no more than a phase of that movement—can that man step out of it and say to himself: I am a fragment of a spiral which is developing in such and such a way? Now it is certain—and more-over Teilhard de Chardin himself recognizes this—that man is able to judge of his own state. Indeed he knows his own rank amongst

39 *Man's Place in Nature*, pp. 62-63.

the other earthly creatures, and he is even the only one to know objectively both himself and the world. Far from being a mere phase in an indefinite evolution, man essentially represents a central possibility, and one that is thus unique, irreplaceable, and definitive. If the human species had to evolve towards another more perfect and more "spiritual" form, man would not already now be the "point of intersection" of the Divine Spirit with the earthly plane; he would neither be capable of salvation, nor able intellectually to surmount the flux of becoming. To express these thoughts according to the perspective of the Gospels: would God have become man if the form of man were not virtually "god on earth," in other words, qualitatively central as well as definitive with regard to his own cosmic level?

As a symptom of our time, Teilhardism is comparable to one of those cracks that are due to the very solidification of the mental carapace,[40] and that do not open upward, toward the heaven of real and transcendent unity, but downward toward the realm of lower psychism. Weary of its own discontinuous vision of the world, the materialist mind lets itself slide toward a false continuity or unity, toward a pseudo-spiritual intoxication, of which this falsified and materialized faith—or this sublimated materialism—that we have just described marks a phase of particular significance.

---

40 René Guénon, *The Reign of Quantity and the Signs of the Times* (Sophia Perennis, Ghent, New York, 2001), Chapter 15, "The Illusion of Ordinary Life."

4

Religion and Science

Lord Northbourne

When this lecture* became inevitable, I decided, perhaps rather rashly, that I would try to set out what I believe to be the essential factors in a very comprehensive and complicated question: that of the relation between religion and modern science.

I am going to try to outline a situation chiefly marked by an unprecedented intellectual confusion arising out of the fact that the astonishingly rapid advance of modern science has caused many beliefs, axioms and assumptions of very long standing to be seriously questioned. The origins and nature of the universe and the situation of man in it have become matters of doubt and of speculation; such indeed are the very questions to which religion and science appear to give different answers. Now these are not questions of interest only to a few philosophers and theologians, they are of immense and immediate practical importance, simply because everyone, even if he hardly ever thinks at all, acts in accordance with some assumption or other concerning the basic realities of his situation. That assumption dictates the tendency, and therefore the ultimate effect, of all that he does, and if it is false his best endeavors are bound to go astray; and this applies with every bit as much force to the collectivity as to the individual. But in these days, in which there is no established traditional order, no unquestioned hierarchy of the intelligence or of anything else, all fundamental decisions are thrown back on to the judgement of the individual, and few indeed are those who are equipped to stand the strain.

First I must indicate as briefly as possible, what I mean by the words "religion" and "science." Not what other people mean, but

* *Editor's note:* The Fellowship Lecture delivered at Wye College in October, 1965.

what I mean; and I must ask you to try to remember, because I shall not qualify them every time they occur.

The Latin root of the word "religion" is connected with the idea of "binding" or "attachment." First, a very broad definition: religion is the link by which humanity is effectively attached to what is greater than itself. By "humanity" I mean mankind as a whole, past, present and future, with all its achievements, aspirations and potentialities both individual and collective. By the word "greater" I mean "eminently" or "incommensurably" greater. If no such attachment is possible, the word "religion" is superfluous. If it is possible, we ignore that possibility at our peril.

But that broad definition must be narrowed down a little. I am thinking, and I expect most of you (perhaps not all) to be thinking, primarily of the Christian religion. But I cannot include everything that claims to be Christian, for the epithet is used to bolster up all kinds of misconceptions, fantasies and sentimentalities. I do not exclude, with similar reservations, any of what are usually known as the "great religions" of the world. They are defined by the fact that they gave rise to great civilizations; it is therefore presumptuous to suppose that they fail to conform to my first definition, despite obvious differences in their outward forms. It is men and times that differ; religion insofar as it is a human institution differs with them, but in its essentials it is always the same.[1] I specifically exclude the many pseudo-religions of relatively recent origin that have attracted so many adherents and done so much to obscure the essentials of religion properly so called.

By the word "science" I mean the whole body of modern observational science in all its branches, but more especially the philosophy that has grown up round it as distinct from its method. That philosophy has permeated modern civilization, and it governs the thoughts and actions of many people to whom the word "philosophy" means almost nothing. The outlook peculiar to it is now pre-

---

1 The implications of any other view seem to me to be unacceptable. I am, however, far from suggesting that it is wrong to regard a particular religion as the best, or even as the only true religion, in a given set of circumstances individual or collective; on the contrary, it is normal and right, I can do no more than make these assertions, since a dissertation on comparative religion is out of the question here. This aspect of the matter is not vital to my argument. See also the following footnotes.

dominant, and this is something new; it is incontestable that in earlier ages an outlook that can broadly be called "religious" was predominant. Some would prefer to say "superstitious," but that is begging a very vital question. Others might prefer the more general word "traditional."

Is there a conflict between religion and science, and what is its nature if it exists? One can say that certainly there ought not to be a conflict, for each claims both to present truth and to be seeking it, so that the more nearly each justifies its claim the more nearly should they come together; but they don't seem to. I want if I can to indicate how far this is due to the fact that both religion and science have got themselves into a false position, though in very different ways, and how far it is due to fundamental divergences.

Religion and science both claim to be true, and I assert without fear of contradiction in this hall that nothing matters in the end but truth. The human faculty concerned in the appreciation of truth is the intelligence, and intelligence is therefore the highest human faculty. Now intelligence is more than reason alone, for reason must have material to work on; reason is that part of the intelligence which relates one datum to another. The source of the data available to reason is not solely external; in fact it is much more "how we see things" than "what we see." I shall return to this point, which is crucial. Meanwhile the point is that, if religion is true, it must engage the intelligence, and the intelligence above all, even before it engages the will and the emotions. I cannot emphasize this too strongly, particularly because the common assumption seems to be that science has a sort of monopoly of intelligence, and that religion is primarily concerned with the will and the emotions. Science, on its part is not worthy of the name unless it takes into account everything that can come within the range of the intelligence and not one aspect of reality alone.

What then is the universe? The common reaction to that question is to the effect that it cannot at present be answered fully, but that anyhow the only way to find out what the universe is, is by looking at it. The difficulty is that looking at the universe, or at any part of it, can never tell us what it is, but only what it looks like to us. The image is not independent of ourselves who make it. We paint a picture of the universe; it is inevitably highly selective because the material available is limitless, and incidentally includes ourselves. So we choose what interests us, and we also choose the light in which it is to be represented. As with all pictures, the result is more than anything else a picture of our own outlook, however "representational" of the outer world we believe it to be.

Furthermore: the seer is not what he sees. This duality is inherent in the act of observation, to whatever that act may be applied; it defines the act. Each one of you can observe the psycho-physical complex of which his body is the material aspect; therefore that complex is other than the observer, other than yourself. So if anyone thinks either that he as observer is aware of anything but the reflected image of the outside world in himself, or that he as observer can turn round and discover by observation what he himself is, he is in manifest error. Yet if he does not know what he himself is, he cannot possibly understand the nature of the images that constitute his knowledge of the universe.

This is the inescapable dilemma sometimes summed up in the words "the eye cannot see itself." Directly we put ourselves into the position of observers, we elude our own observation. Our relationship to our environment is therefore not as simple as we like to think, for we are part of the universe and cannot separate ourselves from it. If we think we can, we fool ourselves. A common and natural reaction to this would be: "So what? We cannot alter that situation; we have nothing to go on but our powers of observation and deduction, and must do our best with what we have, so why bother our heads with such matters?" The answer is that I am talking about a philosophy of science that dominates the world, and these considerations are fundamental to that philosophy, whether it likes it or not.

Is there anything, then, that we can say for sure about the universe? At least we can say that it is an order, a "cosmos"; it is not a "chaos." The living being is also an order, an organism, a "microcosm"; like the universe it is a whole coordinated by something. What is it that makes the universe what it is, and us what we are, and gives to each its inward unity? This is the goal of philosophy, whether it be based on religion or on science.

Science seeks this coordinating principle in the observable. From this point of view the universe consists of identifiable and numerable entities; it does not matter what you call them, because all terms such as "particles" or "forces" are provisional and analogical, since the ultimate constituents, as at present envisaged, can only be described in mathematical terms. The point is that the nature of those constituents is regarded as being deducible from observation, and further, since they are the fundamental constituents of the universe, the coordinating principle is regarded as being inherent in their nature. Therefore the task of science is to elucidate that nature; and it is assumed that if this could be done, everything would be explained; and "everything" must include the psychic element we can observe in living beings. However, that psychic element comes late into the picture, since living beings are regarded as a late (and possibly rather rare and freakish) development in the evolutionary process; nobody supposes that it is they who arranged the stars. But if we, in the name of science, reject all that is not in principle observable, and regard life as a late evolutionary development, we are forced to assume that these inanimate elementary entities or forces, known or as yet unknown, are so constituted as to have here and there combined and arranged themselves in incredibly complex and relatively stable patterns, in such a way that all the phenomena of life are manifested: not only birth, growth, reproduction and death, but also a consciousness both objective and subjective, an active will, memory, emotion and intelligence itself.

This sounds like nonsense, as indeed it is. Nonsense is the only possible result of any attempt to find the coordinating principle of the observable in the observable, or, what amounts to the same thing, of the relative in the relative. Such attempts can only lead to a going round and round in circles, in search of something that is always round the corner and always will be; to a wrapping up of the mystery—or the miracle—of existence and of intelligence in words that get nowhere, in a desperate endeavor to escape at all costs from mystery and from miracle. But in vain, for this mystery is the only thing from which there is no escape save by death. It is the mystery of our own existence and our own intelligence, at once self-evident and inexplicable.

I must explain in parenthesis that the word "mystery," in its debased and commonplace sense, signifies merely anything that is unknown but in principle discoverable. I use it throughout in its original and proper sense, in which it signifies whatever is too exalted or too comprehensive to be grasped or defined distinctively, though it can in principle be apprehended directly. The mysteries of religion are always of this latter nature: the mysteries of science are of the former.

The very principle of the scientific method is to objectify as far as possible. It uses the intelligence but takes its existence for granted; very practical, very sensible, since for most of the work of the world it is superfluous to do otherwise. But if you by-pass the subject, without which there is not objective knowledge, you must not philosophize.

I am far from suggesting that, because they are not "properties of matter" or anything of the kind, life and love, beauty and joy, and intelligence itself, are not of the stuff of which the cosmos is made. Of course they are; they are inherent in its very cause, in its eternal principle, where they subsist as imperishable possibilities. We are aware only of their manifestation under terrestrial conditions, and that manifestation implies the coexistence of their negation;[2] but they are doubtless manifested under endless other conditions of which we can have no inkling while we cannot see beyond our present state. For our universe, in its totality, only represents one of an indefinite multitude of systems of "compossibles," and we only know or can ever know an insignificant fraction even of our own universe, which in its totality is far more extensive, more varied and more wonderful than the wildest dreams of science could ever make it out to be,[3] as Shakespeare knew well: "there are more things in heaven and earth than are dreamt of in your philosophies."

I said that it is nonsense to try to find the coordinating principle of the relative in the relative. It is in fact completely illogical, if words mean anything. I go one further step and suggest to you that what we are always in reality looking for, what we lean on and what we thirst for, whether we know it or not and whatever we think we are looking for, is in fact the non-relative, that is to say, the Absolute, although it is inherently mysterious, unseizable and non-observable. For instance: if you assert that everything is relative, your state-

---

2 Existence is by derivation a "standing apart." Anything that exists stands apart distinctively from everything else, including its own cause and its own opposite or correlative. Its existence therefore implies that of its opposite or correlative; neither light nor darkness has any distinctive existence without the other. In the light of a real grasp of all that this implies, many puzzling questions sort themselves out.

3 *Editor's note:* ". . . let us return for a moment to the modern scientific outlook, since it plays so decisive a part in the modern mentality. There seems to be absolutely no reason for going into raptures about space-flights; the saints in their ecstasies climb infinitely higher, and these words are used in no allegorical sense, but in a perfectly concrete sense that could be called 'scientific' or 'exact.'

ment is itself relative, that is to say, contingent and mobile. It may be right today and wrong tomorrow, and is scarcely worth making. If you maintain that anything (your statement for instance) is less relative than something else, you are bringing in the Absolute. You may argue that there are regions of relative stability, or modes of higher probability, and that your statement is related to them, and so can be said to be more valid than other statements. What can "relative stability" or "higher probability" be taken to mean? They can only mean "near to something yet more stable" or "still less relative," and so on; and in the end inescapably "near to the unchanging, to the non-relative," that is to say: "nearer to the Absolute."

In fact thought is impossible, it is completely chaotic, save in relation to *the* Absolute unqualified and unqualifiable. We are in fact usually thinking of something "relatively absolute," that is to say, of something that represents the Absolute on a particular plane or in a particular region, rather than of the Absolute itself, but this does not alter the fact that the Absolute constitutes the basic condition and the fundamental assumption of all logical and coherent thought. It is limitless and all-comprehending and therefore undefinable, nevertheless it forces itself upon us even when we ignore it or try to dispense with it. If we try to escape from it, we inevitably end up by inventing a false absolute, which amounts to adopting an unreal and invalid point of reference. This fact is by no means unconnected with the fact that if we try to dispense with God we inevitably end by inventing false gods; and this is true although the word "absolute" and the word "God" are not interchangeable. And when false gods fail it, as they must, humanity has nothing left to

In vain does modern science explore the infinitely distant and the infinitely small; it can reach in its own way the world of galaxies and that of molecules, but it is unaware—since it believes neither in Revelation nor in pure intellection—of all the immaterial and supra-sensorial worlds that as it were envelop our sensorial dimensions, and in relation to which these dimensions are no more than a sort of fragile coagulation, destined to disappear when its time comes before the blinding power of the Divine Reality. To postulate a science without metaphysic is a flagrant contradiction, for without metaphysic there can be no standards and no criteria, no intelligence able to penetrate, contemplate and coordinate. Both a relativistic psychologism which ignores the absolute, and also evolutionism which is absurd because contradictory (since the greater cannot come from the less) can be explained only by this exclusion of what is essential and total in intelligence." Frithjof Schuon, *Light on the Ancient Worlds* (World Wisdom, Bloomington, Indiana, 1984), p. 130.

deify but itself. This development has a name: "humanism" we call it.

The rightful domain of science is that of the observable, and surely it ought to be enough, for it is inexhaustible, though so very far from being everything. The rightful domain of religion is that of the fundamental but non-observable mystery, call it what you will, that is the key to everything, though some who claim to represent religion seem to behave as if they had forgotten the fact. Conflict and confusion arise when either tries to occupy the domain of the other. Science gets into trouble and ends up nowhere when it tries to philosophize about ultimates, instead of getting on with its entirely practical work, its craft. Religion gets into trouble when it tries to adapt itself to the approach of science, instead of trying to perfect its own approach.

We are obsessed by the fact that we have found out how to do so much to enlarge the sensitivity of our organs of sense, by the use of telescopes, microscopes and all the rest. We forget that it is what we are, our own inmost nature, the "light that is in us" that conditions what we make of the messages we receive through the senses, and that is vastly more important than how many different sense-impressions we receive; "And if that light be darkness, how great is that darkness."[4] We forget that a mere multiplication of facts (which is, in the strictest sense of the word, interminable) can do nothing whatever towards improving the quality of our intelligence; indeed, when it becomes an obsession, it can easily lead to a fragmentation of knowledge rather than to its unification. I would go farther, and say that it inevitably does so; also that computers cannot help, because they are not intelligent. The most widely traveled individual is not necessarily the wisest; a hermit may be far wiser than he. It is indeed perfectly possible to see too much, and, in the common phrase, to be unable to see the wood for the trees. It is equally possible to look so hard in one direction that you see nothing in the other; to be so preoccupied with your botanizing that you do not notice the bull charging you from behind.

Only one who knows what his own existence is (and he cannot find out by observation, nor can he know any existence but his own) knows what existence as such is, that of other people and things, as

4 Matt. 6:23.

well as his own.[5] Not how he himself or other people or things look or behave, that can be learnt by observation, but what they are, what it is that behaves in such and such a way, whether its appearance be that of a man or an atom or a star. It is ten thousand times more important to know what man is, even imperfectly, than to know, however completely, the distances of the stars or how to smash an atom. It is perhaps not surprising that this kind of knowledge is often most accessible, intuitively but not analytically, to the mentally uncomplicated, and is "hidden from the wise and prudent."[6] You may recall too that the "mystery of the Kingdom of God . . . cometh not with observation" but is "within you."[7]

Not for nothing was the inscription "Know Thyself" written over the gateway to Aristotle's school of philosophy; but of course his philosophy was founded on religion. Religion is there to teach us what we are—each according to his capacity to accept and to understand—and, insofar as it does so, not only does it engage the intelligence, but it is the very foundation of intelligence. "To fear the Lord is the beginning of wisdom," said Ecclesiasticus.[8]

Someone may be thinking: "What is all this preoccupation with oneself? Surely it is contrary to religion as well as to our natural feelings, and surely the one thing that is useful and unselfish is to get

---

5 *Editor's note:* "Modern science, which is rationalist as to its subject and materialist as to its object, can describe our situation physically and approximately, but it can tell us nothing about our extra-spatial situation in the total and real Universe. Astronomers know more or less where we are in space, in what relative 'place,' in which of the peripheral arms of the Milky Way, and they may perhaps know where the Milky Way is situated among the other assemblages of stardust; but they do not know where we are in existential 'space,' namely, in a state of hardness and at the center or summit thereof, and that we are simultaneously on the edge of an immense 'rotation,' which is not other than the current of forms, the 'samsaric' flow of phenomena, the *panta rhei* of Heraclitus. Profane science, in seeking to pierce to its depth the mystery of the things that contain—space, time, matter, energy—forget the mystery of the things that are contained: it tries to explain the quintessential properties of our bodies and the intimate functioning of our souls, but it does not know what intelligence and existence are; consequently, seeing what its 'principles' are, it cannot be otherwise than ignorant of what man is." Frithjof Schuon, *Light on the Ancient Worlds* (World Wisdom, Bloomington, Indiana, 1984), p. 111.

6 Matt. 11:25.

7 Luke 17:20 and 21.

8 Ecclesiasticus 1:14.

on as best we can with making the world a better place; for we can only take things as we find them, ourselves included, and do our best with them."

There are two immediate answers. Firstly: action cannot be effective unless based on a knowledge as accurate and as comprehensive as possible. Goodwill is of course necessary, but, alone it is powerless. If what I have said is right, and if the key to understanding is in the answer to the question, "What am I?" which cannot be answered by observation, then to seek it where it is to be found, namely "within you," cannot be selfish; even apart from the fact that no task is more exacting than that search, which necessitates (at first sight paradoxically) the elimination of all personal ambition or desire. Nor is any task more charitable, since its fulfillment alone can teach us what we are. As a specialist task it is by no means everyone's: it demands both vocation and training; but all other tasks are justified by the extent to which they help to make it possible. This may seem a surprising assertion, yet that is the principle underlying the structure of every civilization founded on religion, however imperfectly it may be realized. No wonder we don't understand such civilizations.

Secondly: the objective of action must be clear and valid. It cannot be either if it is based on uncertainty or misconception about what man is, or about what are his origin, function and destiny. Where any such misconception exists, efforts to do good are likely to be misdirected. That is putting it mildly. "Where there is no vision the people perish."[9]

Most of our actions today are dictated by a combination between a philosophy of science, more or less popularized, and habits of thought originating in a religion that has largely lost its original authority. I am evidently implying that this combination is weak in its understanding of the origin, function and destiny of man. That there should be confusion is not surprising, for it is in their respective views of man's situation that religion and science differ most conspicuously. In discussing their differences, I shall of course use the religious language that is familiar to most of us. It is as adequate as words can be for giving expression to ideas concerning the mystery of existence but it is essential not to forget that it is sym-

9 Proverbs 29:18.

bolical, because it cannot be "descriptive" in the limitative sense of the word.[10]

Let us consider origin and function first. According to the religious view, the origin of all things is divine, and therefore mysterious in the proper sense. Man is the culminating point of the creation, the representative of God on earth, and his special function is to keep the universe in touch with God, who is its origin and its end, and this implies that he must above all keep himself in touch with God. For this purpose he was created and has been given his dominion over the animals and plants. But let me quote St. Francis of Sales (not of Assisi) who, in his *Introduction to the Devout Life*, puts the religious view of the function of man in its purest— some would say its most extreme—form.

> God did not put you into this world because of any need that he had of you, but only that he might exercise in you his goodness, giving you his grace and his glory. To this end he has given you understanding wherewith to know him, memory wherewith to remember him, will wherewith to love him, imagination that you

10 Not all religions envisage the origin of the universe in terms of a divine "Creation," as do Judaism, Christianity, and Islam. Not all are even theistic, for Buddhism is not. The mystery that underlies all existence can be symbolized in many different ways, not necessarily outwardly coincident, much in the same way as separate two-dimensional projections of a solid object may differ according to the point of view without being intrinsically false. They may suggest the third dimension, but cannot specify it by their form alone. We rightly seek precision in our statements, but a statement can be precise in two senses: either because it is inherently unequivocal, or because it is understood as it was intended to be. Only one kind of statement inherently is unequivocal, and that is the purely quantitative, of which the type is "two and two make four" and the development is constituted by mathematical formulae of all degrees of complexity. Quantity by itself has no significance, however elaborate its formulation; in order to be significant it must be related to something qualitatively distinguishable. In our efforts to obtain precision we are continually seeking to reduce quality to quantity, that is to say, to reduce reality to mathematical formulae. The result is that the great positive qualities: love, beauty, goodness, mercy, intelligence and so on, are relegated to a secondary position, as if they were purely human and subjective, whereas in reality they lie at the heart of everything. For the world, inanimate as well as animate, is constituted by the interaction of quality and quantity, which very broadly correspond to what we call "spirit" and "material" respectively. In trying to express everything in terms appropriate to the "material" aspect alone we lose sight of the spirit. A statement having a qualitative significance can be perfectly precise, despite the fact that the possibility of misunderstanding cannot be eliminated.

might picture his benefits, eyes that you might see the marvels of his works, a tongue wherewith to praise him, and likewise with the other faculties. Being created and put into this world with that intention, all intentions contrary thereto must be rejected and avoided, and those that in no way serve this end must be despised as being vain and superfluous. Consider the misfortune of the world which thinks not at all of this, but lives as if thinking that it had been created only to build houses, plant trees, amass riches and disport itself.

The scientific view, in its purest or most extreme form, is that all things are the product of an evolutionary process, the details of which it is the task of science to elucidate. They are mysterious only in the popular sense. According to his view, man is a product of evolution; his faculties have been developed step by step, by a process not yet fully understood, but in principle ascertainable. The function of man is therefore whatever he likes to make it, and in practice, to look after himself. If he has a responsibility towards his neighbors, human and non-human, it is a matter of conscience or of mutual advantage; and conscience itself must be a product of evolution; and evolution cannot be allowed by many advanced contemporary philosophers to be purposive in any sense, for fear of admitting the idea, however attenuated, of a god of some sort. So the best that man can do is to derive as much advantage to himself as he can from the accidents of his constitution and of his environment.

I cannot begin to see how these two points of view can be reconciled, unless they are so watered down as to be unrecognizable. However they may be formulated, the priorities implied by each are diametrically opposed.

So much for origin and function. What about destiny? Or if you like a nice scientific-sounding word: eschatology? Religion says that God gave us our lives and that at death they are taken back by Him. Our bodies are but the temporary dwelling-place of an immortal soul, which is subject to a judgement after death, as a result of which it goes to paradise, purgatory or hell. This aspect of religion is often nowadays glossed over as far as possible, but that does not alter the fact that it is absolutely essential.[11]

---

11 Our eschatological situation is beyond the reach of our imagination, which is derived entirely from our terrestrial experience. The symbolical image of it that

It may occur to you that if what you the observer, the knower, the subject really are is other than the psycho-physical complex you can observe, there is no particular reason why you should perish when it perishes. But the exclusively scientific mind shies at such ideas because they cannot be checked in any way by observation. How could they? They are concerned only with that inmost "I" which we cannot observe, but which is nevertheless our real selves, on the one hand, and on the other with a state in which the real self is detached from the conditions of its mundane existence, including time and space. And if anyone says that only things tied to time, space or other terrestrial conditions can have any relevance for us or contact with us during our lives, I reply that it is precisely the intelligence that is not tied in that way—unless indeed it insists on forging its own chains.

By contrast, the eschatology of observational science is extremely simple, for the method it employs can never reveal any reason for regarding death—the only certainty that faces all of us— as anything but a total extinction; indeed if man is identified with his body it can be nothing else. (I might mention by the way, that there is no need to question the reality of some of the phenomena associated with spiritualism; the interpretation to be assigned to them is a very different matter.) If extinction be in fact our destiny, the hitherto almost universal belief of humanity in some sort of "after-life" must be a delusion, no doubt largely wishful in origin, and must be replaced as quickly as possible by a more realistic view. A realistic view must however take account of every aspect of reality inward as well as outward: so which view is really realistic? And incidentally which is really dispassionate? It seems to me that the postulate of total extinction can serve as an easy way out of the necessity of facing the dread alternative of a heaven and a hell, and the prospect of a judgement in which our smallest and least considered actions and attitudes may outweigh all those we now regard as significant, because it is they that give the show away. And then, with the veil of the flesh torn away, at last we really see ourselves.

An eschatological compromise seems even more impossible than in the cases of origin and function. Either religion is childish and misleading, and destined to give way to an intellectual maturity

most adequately suggests its reality to a particular sector of humanity is the image presented by the religion characteristic of that sector.

of very recent appearance and great potentiality; or else science, insofar as it concerns itself with the origin of the universe or the function and destiny of man, is just plain wrong.

Such fundamental divergences impose an ineluctable choice. I have suggested that, since truth is in question, that choice must be referred to the intelligence, bearing in mind that intelligence is more than reason alone. I should not be surprised if some of you are thinking that in that connection I am attributing more to religion than is really there. More than meets the eye of the casual or unsympathetic observer—yes; more than it is easy for the unprejudiced but puzzled enquirer to find—perhaps; but more than is there—no.

The enemies of religion are interested above all in making it appear to be as arbitrary, as non-essential and as unintellectual as possible. One would sometimes think that some of its defenders, in their efforts to popularize it, were prepared to go a very long way in the same direction. I have made frequent reservations concerning religion in connection with some of its contemporary tendencies, all pointing to the fact that its intellectual aspect—the doctrinal aspect that engages the intelligence and is "metaphysical" in the proper sense of that much abused word,[12] or "philosophical" in the ancient sense of that word—that aspect has become so obscured by an overlay of moralism and emotionalism as almost to be forgotten. Nevertheless, it is always present, and accessible to those that "have ears to hear," in the words of the sacred Scriptures and of their orthodox commentaries; it is also implicit in the outward forms of religion, including its doctrinal formulations and its ritual which, if they had no intellectual basis, would indeed be arbitrary. This intellectual or metaphysical background is the heart of all religion, and, when it is lost sight of, religion cannot but go astray.

---

12 The word "metaphysical" comes from the Greek. It does not mean "beyond the physical" in the current sense of the last word, but rather "beyond the natural," that is to say "beyond the observable." It is therefore equivalent to the Latin "supernatural," provided that the latter is understood literally and not in its degraded sense, in which it is applied to almost any unexplained phenomenon. Properly speaking, neither word is concerned with phenomena as such, but exclusively with the universal principles underlying all phenomena, explicable or otherwise; and that is as much as to say—with the "mystery" in the ancient sense (from a Greek word meaning "to be silent"). Therefore the language of metaphysic is always symbolical and not descriptive; it must leave room for the inexpressible.

There are two ways of accounting for the hold that religion has maintained on mankind since the dawn of history until now—or should I say—"until very recently"? One is that there appears to be a kind of religious phase, factual but difficult to explain, in the evolutionary progress of humanity from a relatively bestial state to a civilized maturity, of which the present age is probably only the beginning.

The other is related to what I have just said, that the ultimate truth about the nature of the universe and the situation of man is implicit in, and somehow shines through, the very varied forms of religion; and that it is the concordance of this truth with our own inmost nature that confers on religion its mysterious power to attract and to hold. This truth is too comprehensive to be contained by any unequivocal dialectical formulation, so that, for a large majority at least, religious conformity in the shape of belief and observance brings them much nearer to the truth than anything else possibly could. Religious belief therefore is a manifestation of intelligence, at least insofar as it is the expression of a real inward understanding which is unable to express itself in any other way, and moreover has no need to try to do so. Religion takes man as he is, and not as if everyone were a saint or a sage.

Belief is the form in which religious truth reaches the many. There are always some whom it reaches in a more explicitly intellectual form, and they alone are qualified to oppose dialectically any system of ideas that contradicts either religion as a whole or a particular religion. When those who are sufficiently well qualified are too few, or when pandemonium prevents their voices from being heard, religion is led into making more and more compromises, not with facts, which it never denies, but with a philosophy which seeks to reduce God to the measure of man, even when it does not reject God flatly, in either case depriving man of the possibility of rising above himself. The real strength of religion lies in its conformity to its metaphysical background,[13] in the light of which a synthetic view of the complexities of experience becomes

---

13 The essential unity of the great religions resides in their conformity to this common metaphysical background, and in nothing else. That background has been called the "*philosophia perennis*"; it is the "undying wisdom" that is the heritage of the whole human race.

possible, and in which the situation of man becomes clear in all its essentials. The strength of religion lies also, humanly speaking, in the uncompromising nature of its doctrines, provided that it does not admit compromise.

There can be no justifiable criticism of the precision of science, nor of its objectivity, its quasi-mathematical detachment and (in theory at least) its dispassion. The effectiveness of your work depends on your maintaining those characteristics to the utmost; but their inherent limitation must be recognized. A good thing can get out of place, and I am certainly not suggesting that it is the fault of you, the practical men, that it has happened in this case. The fact is that the approach of science does not get to the heart of things, and it is impossible that it ever should. Nobody denies its effectiveness in changing the face of the world, and in providing us with material possessions on a scale hitherto undreamt of, and in combating disease and pain. Nevertheless, its application has not yet produced much contentment or feeling of security, which seems to be as far off as ever, if not farther. Why do people persist in their quarrels and discontents, hatreds, suspicions and revolts, and show no signs of amendment? Is it really because they have not yet got enough, or because someone else's lack of goodwill or stupidity delays the raising of the standard of living everywhere? It becomes daily more difficult to make that kind of explanation fit the facts. Or is it in the last analysis because even those who are most abundantly equipped for living are starved as never before of all that could give meaning to their lives, and because what is being offered to them—or should I say: what is being thrust down their throats?—does nothing whatever towards meeting this, the first of all needs?

If that be so, I suggest that the reason is that we, whatever may be our *credo*, have in practice behaved as if this life carried its own justification in itself, and have chosen to treat our existence as if it were an accident, and our intelligence as no more than a tool for the satisfaction of earthly needs and desires; whereas in reality that intelligence, provided that we are not too proud to acknowledge the mystery of its origin and of our own, can penetrate beyond the confines of the universe of phenomena and give us a glimpse of what is greater than ourselves; and that is what we need above all to give direction and meaning to our lives, to give us something to live for, and something to die for.

5

Contemporary Man,
between the Rim and the Axis[1]

Seyyed Hossein Nasr

Come you lost Atoms to your Center draw,
And be the Eternal Mirror that you saw:
Rays that have wander'd into Darkness wide
Return, and back into your Sun subside.

Farîd al-Dîn 'Attâr

My Guru taught me but one precept. He told me,
"From the outward enter unto the most inward
part of thy being." That has become for me a rule.

Lallâ, the Female Saint of Kashmir, *Lallâ Vakyanî*, 94

The confrontation between man's own inventions and manipulations in the form of technology and human culture as well as the violent effect of the application of man's acquired knowledge of nature to the destruction of the natural environment have reached such proportions that many people in the modern world are at last beginning to question the validity of the conception of man held in the West since the rise of modern civilization. But to discuss such a vast problem in a meaningful and constructive way one cannot but begin by clearing the ground of the obstacles which usually prevent the profoundest questions involved from being discussed. Modern man has burned his hands in the fire which he himself has kindled by allowing himself to forget who he is. Having sold his soul, in the manner of Faust, in order to gain dominion over the natural environment, he has created a situation in which the very control of the environment is turning into its strangulation,

---

1 This essay contains the text of the Charles Strong Memorial Lecture which the author delivered in Austria in 1970.

bringing in its wake not only ecocide but also ultimately suicide. The danger is now evident enough not to need repetition. Whereas only two decades ago everyone spoke of man's unlimited possibility for development, understood in a physical and materialistic sense, today one speaks of "limits to growth"—a title well-known to many people in the West today—or even of an imminent apocalypse. But the concepts and factors according to which the crisis is analyzed, the solutions sought after and even the colors with which the image of an impending doom are depicted are all in terms of the very elements that have brought the crisis of modern man into being. The world is still seen as devoid of a spiritual horizon, not because there is no such horizon present but because the subject who views the contemporary landscape is most often the type of man who lives at the rim of the circle of existence and therefore views all things from the periphery. He remains indifferent to the spokes and completely oblivious to the axis or the Center which remains ever accessible to him through the spokes of the wheel of existence.

The problem of the devastation brought upon the environment by technology, the ecological crisis and the like all issue from the malady of *amnesis* or forgetfulness from which modern man suffers. Modern man has simply forgotten who he is. Living on the periphery of his own circle of existence he has been able to gain a qualitatively superficial but quantitatively staggering knowledge of the world. He has projected the externalized and superficial image of himself upon the world.[2] And then, having come to know the world in such externalized terms he has sought to reconstruct an image of himself based upon this external knowledge. There has been a series of "falls" by means of which man has oscillated in a descending scale between an ever more externalized image of

---

2 It must be remembered that, in the West, first man rebelled against Heaven with the humanism of the Renaissance and only later did the modem sciences come into being. The humanistic anthropology of the Renaissance was a necessary background for the scientific revolution of the 17th century and the creation of a science which although in one sense non-human is in another sense the most anthropomorphic form of knowledge possible, for it makes human reason and the empirical data based upon the human senses the sole criteria for the validity of all knowledge.

Concerning the gradual disfiguration of the image of man in the West see G. Durand, *"Défiguration philosophique et figure traditionelle de l'homme en Occident,"* *Eranos-Jahrbuch*, XXXVIII, 1969, pp. 45-93.

himself and of the world surrounding him, moving ever further from the Center of both himself and of his cosmic environment. The inner history of the so-called development of modern man from his historic background as traditional man, who is at once his ancestor in time and his center in space, is a gradual alienation from the center and the axis through the spokes of the wheel of existence to the rim wherein resides modern man. But just as every rim presupposes spokes which connect it to the axis of the wheel, so does the very fact of human existence imply the presence of the center and the axis and hence an inevitable connection of men of all ages with Man as such, with the *anthropos*, or *al-insân al-kâmil* of Sufism, as he has been, is and will continue to be, above all outward changes and transformations.[3]

Nowhere is the attempt to solve the problems caused by the various activities of modern man by refusing to consider the negative nature of the very factors that have caused these problems more evident than in the field of the humanities in general and the sciences dealing specifically with man, which are supposed to provide an insight into human nature in particular. Modern man, having rebelled against heaven, created a science based not on the light of the intellect but on the powers of human reason sifting the data of the senses. But the success of this science was so great in its own domain that soon all the other sciences began to ape it, leading to the crass positivism of the past century which has caused philosophy as perennially understood to become confused with logical analysis, mental acrobatics or even mere information theory, and the classical fields of the humanities to become converted into quantified social sciences which make even the intuitions of literature about the nature of man inaccessible to many students and seekers today. A number of scientists are in fact among those most critical of the pseudo-humanities being taught in many universities in an atmosphere of a psychological and mental inferiority complex vis-à-vis the

---

3 If such a relation did not exist, it would not even be possible for man to identify himself with other periods of human history, much less for the permanent aspects of human nature to manifest themselves even in the modern world as they have in the past and continue to do so today.

sciences of nature and mathematics, a "humanities" which tries desperately to become "scientific," only to fall into a state of superficiality, not to say triviality.[4] The decadence of the humanities in modern times is caused by the loss of the knowledge that man has always had directly of himself and also of his Self, and by reliance upon an externalized, indirect knowledge which modern man seeks to gain of himself from the outside, a literally "superficial" knowledge that is from the rim and is devoid of an awareness of the axis of the wheel and the spokes which stand always before man and connect him like a ray of light to the supernal sun.

It is with consideration of this background that certain specific questions which come to mind must be analyzed and answered. The first query concerns the relation of small pieces of scientific evidence about human behavior to "human nature." In order to answer this question it is essential to remember that the reality of the human state cannot be exhausted by any of its outward projections. A particular human action or behavior always reflects a state of being, and its study can lead to a certain kind of knowledge of the state of being of the agent, provided there is already an awareness of the whole to which the fragment can be related. Fragmented knowledge of human behavior is related to human nature in the same way that waves of the sea are related to the sea. There is certainly a relationship between them that is both causal and substantial. But unless one has had a vision of the sea in its vastness and illimitable horizons, which reflect the Infinite and its inimitable peace and calm, one cannot gain an essential knowledge of the sea through the study of the waves. Fragmented knowledge can be related to the whole only when there is already an intellectual vision of the whole.

The careful "scientific" study of fragmented human behavior is incapable of revealing the profounder aspect of human nature precisely because of an a priori limitation that so much of modern behavioristic studies of man, a veritable conglomerate of pseudo-sci-

---

4 Certain American scholars such as William Arrowsmith have already criticized what could be called the "pollution of the humanities," but the tendency here as in the pollution of the environment is mostly to try to remove the ill effects without curing the underlying causes.

ences if there ever was one,[5] has placed on the meaning of the human state itself. There has never been as little knowledge of man, of the *anthropos,* in different human cultures as one finds among most modern anthropologists today. The medicine men of Africa have had a deeper insight into human nature than the modern behaviorists and their flock, because the former have been concerned with the essential and the latter with accidentals. Now, accidents do possess a reality, but they have a meaning only in relation to the substance which supports them ontologically. Otherwise one could collect accidents and external facts indefinitely without ever reaching the substance or what is essential. The classical error of modern civilization to mistake the quantitative accumulation of information for qualitative penetration into the inner meaning of things applies here as elsewhere. The study of fragmented behavior without a vision of the human nature which is the cause of this behavior cannot itself lead to a knowledge of human nature. It can go around the rim of the wheel indefinitely without even entering upon the spoke to approach the proximity of the axis and the Center. But if the vision is already present, the gaining of knowledge of external human behavior can always be an occasion for re-collection and a return to the cause by means of the external effect. In Islamic metaphysics, four basic qualities are attributed to Ultimate Reality, based directly on the Koranic verse, "He is the First and the Last, the Outward and the Inward" (57: 3). This attribution, besides other levels of meaning, also possesses a meaning that is directly pertinent to the present argument. God, the Ultimate Reality, is both the Inward (*al-Bâtin*) and the Outward (*az-Zâhir*), the Center and the Circumference. The religious man sees God as the Inward; the profane man who has become completely oblivious to the world of the Spirit sees only the Outward, but precisely because of his ignorance of the center does not realize that even the Outward is a

---

5 In modern times the occult sciences, whose metaphysical principles have been forgotten, have become known as the pseudo-sciences, while in reality they contain a profound doctrine concerning the nature of man and the cosmos. Much of the social and human sciences today on the contrary veil and hide a total ignorance of human nature with a scientific garb and are in a sense the reverse of the occult sciences.

manifestation of the Center or of the Divine. Hence his fragmented knowledge remains incapable of seizing the whole of the rim or circumference and therefore by anticipation the Center. A segment of the rim remains nothing more than a figure without a point of reference or Center, but the whole rim cannot but reflect the Center. Finally the sage sees God as both the Inward and the Outward. He is able to relate the fragmented external knowledge to the Center and see in the rim a reflection of the Center. But this he is able to do only because of his a priori awareness of the Center. Before being able to see the external world—be it the physical world about us or the outer crust of the human psyche—as a manifestation of the Inward, one must already have become attached to the Inward through faith and knowledge.[6] Applying this principle, a sage could thus relate fragmented knowledge to the deeper layers of human nature, but for one who has yet to become aware of the Inward dimension within himself and the Universe about him, fragmented knowledge cannot but remain fragmentary, especially if the fragmentary knowledge of human behavior is based upon observation of the behavior of a human collectivity most of whose members themselves live only on the most outward layers of their being and whose behavior only rarely reflects the deeper dimension of their own being.

This last point leads to an observation that complements the discussion of principles already stated. Modern man lives for the most part in a world in which he encounters few people who live on the higher planes of consciousness or the deeper layers of their being. He therefore is for the most part aware of only certain types of human behavior. Fragmented knowledge of human behavior, even if based on external observation, could be an aid for modern man

---

6 This theme is thoroughly analyzed by Frithjof Schuon in his *Form and Substance in the Religions* (World Wisdom, Bloomington, Indiana, 2002). Concerning the sage or the Sufi he writes, "The Sufi lives under the gaze of *al-Awwal* (the First), *al-Âkhir* (the Last), *az-Zâhir* (the Outward) and *al-Bâtin* (the Inward). He lives concretely in these metaphysical dimensions as ordinary creatures move in space and time, and as he himself moves in so far as he is a mortal creature. He is consciously the point of intersection where the Divine dimensions meet; unequivocally engaged in the universal drama, he suffers no illusion about impossible avenues of escape, and he never situates himself in the fallacious 'extra-territoriality' of the profane, who imagine that they can live outside spiritual Reality, the only reality there is." p. 76.

to become at least indirectly aware of other dimensions of human nature, provided a study is made of the behavior of traditional man, of the man who lives in a world with a Center. The behavior of traditional men of different societies, especially at the highest level of the saints and sages, be they from the Chinese or the Islamic or the North American Indian or any other traditional background, in the face of great trials, before death, in presence of the beauty of virgin nature and sacred art, or in the throes of love both human and divine, can certainly provide indications of aspects of human nature for the modern observer. Such behavior can reveal a constancy and permanence of human nature that is truly astonishing and can also be instrumental in depicting the grandeur of human nature, which has become largely forgotten in a world where man has become a prisoner of the pettiness of his own trivial creations and inventions. Seen in this light a fragmented knowledge of human behavior can aid in gaining a knowledge of certain aspects of human nature. But in any case a total knowledge of human nature cannot but be achieved through a knowledge of the Center of the axis, which also "contains" the spokes and the rim. A famous saying of the Prophet of Islam states, "He who knows himself knows his Lord." But precisely because "himself" implies the Self which resides at the Center of man's being, from another point of view this statement can also be reversed and it can be stated that man can know himself completely only in the light of God, for the relative cannot be known save with respect to the Absolute.

The second query to which we must address ourselves concerns the relationship of scientific "objectivity" and its findings to the criteria of "the universal and the unchanging" implied by the phrase "human nature." To answer this query it is necessary before all else to define once again what is meant by scientific "objectivity," especially when it concerns the study of man. It has become commonplace, at least for non-specialists in the philosophy of science, to attribute objectivity to modern science almost as if the one implied the other. No doubt modern science possesses a limited form of "objectivity" in its study of the physical world, but even in this domain this "objectivity" is encompassed by the collective subjectivity, of a particular humanity at a certain moment of its historical existence when the symbolist spirit has become atrophied and the gift of seeing the world of the spirit through and beyond the physical world has been nearly completely lost. Even in the physical world all that cannot be caught in the net of modern science, to quote the well-known image of A. Eddington, is collectively neglected, and its non-existence vowed for "objectively." It is as if an audience of deaf people testified together that they did not hear any music from musicians playing before them and considered the unanimity of their opinion as a proof of its objectivity.

Now if in the domain of the physical world itself the concept of the so-called "objectivity" of modern science must be employed with great prudence and the qualitative and symbolic aspects of nature not neglected because of their lying outside the "objectively" defined world view of modern science, so much more does this "objectivity" need to be re-considered in the field of the study of man. The aping of the methods of the physical sciences in the study of man have enabled scientists to gather a great deal of information about men of all ages and climes but little about man himself, for the simple reasons that the philosophical background of modern science, which goes back ultimately to Cartesianism, is incapable of providing the necessary background for the study of man. Already in the 17th century the body-mind dualism of Descartes perverted in the European mind the image of the much more profound tripartite division of the being of man consisting of corpus, anima and spiritus expounded so fully in the Hermetic tradition. To this error a worse illusion was added in the 19th century which prevented even the collecting of facts about men of different ages from becoming a way of reaching at least some form of knowledge of man himself.

This illusion is that of evolution as it is usually understood today. Evolution is no more than a scientific hypothesis that has been parading around for the past century as a scientific fact, despite the lack of the least amount of proof of its having taken place in the biological plane and despite its being usually taught in schools as proven fact. The present discussion does not allow our entering into debates about biological evolution, although writings by biologists and geologists against it, especially works written during recent years, are far from being few in number.[7] But as far as the study of

---

7 See for example, L. Bounoure, *Déterminisme et finalité, double loi de la vie*, (Paris, 1957) ; his *Recherche d'une doctrine de la vie*, (Paris, 1964) ; and D. Dewar, *The Transformist Illusion*, (Murfreesboro, 1957). See also S. H. Nasr, *The Encounter of Man and Nature, The Spiritual Crisis of Modern Man*, (London, 1968), pp. 124ff. where works and views opposed to evolution are discussed.

man is concerned, it is precisely the intrusion of the idea of evolution into anthropology that has made the positive relation which scientifically accumulated facts could have had to an understanding of the universal and unchanging aspect of "human nature" well nigh impossible. Scientists and scholars in the fields of anthropology, the social sciences and even the humanities are trained almost completely to study only change. Any alteration, no matter how trivial, is more often than not considered as a significant change, while the immutable is almost unconsciously identified with the unimportant or the dead. It is as if man were trained to study only the movement of clouds and to remain completely oblivious of the sky with its immutable and infinite expanses which provides the matrix for the observations of the cloud movements. No wonder that so much of the study of man provided by modern disciplines is really no more than a study of triviality producing most often petty results and failing at almost every step to predict anything of significance in the social order. Many a simple traditional folk tale reveals more about man than thick tomes provided with pages of statistics on what is usually described as "vital changes." In fact the only vital change that is occurring today is the ever greater alienation of man from his own permanent nature and a forgetfulness of this nature, a forgetfulness which cannot but be ephemeral and is bound to have catastrophic effects upon that type of man who has chosen to forget who he is. But this is precisely the one change which "objective" scientific methods are incapable of studying.

Yet, in principle, there is no necessary contradiction between scientific facts accumulated objectively and the concept of "human nature" with its permanent and universal implications. Were the impediments of that mental deformation called evolutionary thinking, which is neither "objective" nor scientific, to be removed, the accumulation of facts about man would display in a blindingly evident fashion the extra-spatial and extra-temporal character of man, if not beyond history—for this would lie beyond the reach of facts—at least in periods of history and in various regions of the world. Such an exercise would depict human nature as something constant and permanent, from which at certain moments of history and among certain peoples there have been deviations and departures that have soon been removed by tragedies or catastrophes leading to a reestablishment of the norm. The sacred books such as the Koran contain, besides other levels of meaning, a "history" of the human soul which emphasizes in a majestic fashion this conception of human nature.[8] That is why the goal that is placed before man in all sacred books is to know and to return to the norm, to man's permanent and original nature, the *fitrah* of the Koran. As the *Tao Te Ching* (XIX) states, "Realize thy Simple Self. Embrace thy Original Nature." For the goal of man cannot be but the knowledge of himself, of who he is.

> He who knows others is wise;
> He who knows himself is enlightened
>
> *Tao Te Ching*, XXXIII

Or to quote a Western contemplative,

> If the mind would fain ascend to the height of science,
> let its first and principal study be to know itself.
>
> Richard of St. Victor

In the light of the understanding which both revelation and intellectual vision have provided over the ages concerning the nature of man, the answer to the oft-posed question "Can we know that any scientific knowledge we may gain captures something *essential* about man?" can only be the following: We cannot gain an essential knowledge of man through any method that is based on an externalization of man's inner being and then the placing of this externalized man, of the man who stands at the rim of the wheel of existence, as the subject that knows. If *essential* has any meaning at all it must be related to the essence, to the center or axis which generates at once the spokes and the rim. Only the higher can comprehend the lower, for to *comprehend* means literally to encompass, and only that which stands on a higher level of existence can encompass that which lies below it. Man is composed of body, psyche and intellect, the latter being at once above man and at the center of his being. The essence of man, that which is *essential* to human nature, can be understood only by the intellect, through the "Eye of the heart" as traditionally understood, the intellect which is

---

8 For the episodes of the Koran considered as events of the human soul and its inner "history" see Frithjof Schuon, *Understanding Islam* (World Wisdom, Bloomington, Indiana, 1998), Chapter 2.

at once at the center of man's being and encompasses all of his other levels of existence. Once the eye of the heart becomes closed and the faculty of intellection, in its original sense, atrophied, it is not possible to gain an essential knowledge of man. The reflection of the intellect upon the plane of the psyche and the mind, which is reason, can never reach the essence of man or for that matter the essence of anything else, no matter how much it concerns itself with experiment and observation or how far it carries out its proper function of division and analysis, the legitimate and rightful function of *ratio*. It can gain peripheral knowledge of accidents, of effects, of external behavior, but not of the essence. Reason, once divorced from the guiding light of the intellect, can at best confirm the existence of the *noumena*, of the essences of things, as we see in the philosophy of Kant, but it cannot know that essence. The knowledge that is essential is one that is ultimately based on the identity of the knower and the known, on being consumed by the fire of knowledge itself. Man is at a particular vantage point to know one thing in essence, and that is himself, were he only to overcome the illusion of taking, to use Vedantic terms, the externalized and objectified image of himself for his real Self, the Self which cannot be externalized because of its very nature. Scientific knowledge, like any other form of knowledge which is based by definition on the distinction between a subject that knows and an object that is known, must of necessity remain content with a knowledge that is peripheral and not essential.

One is naturally led to ask what is the relationship of particular scientific research to the quest for other kinds of knowledge about mankind in general. A relation of a legitimate and meaningful kind can exist provided the correct proportion and relation between ways of knowing is kept in mind. And that is possible only if a knowledge that transcends science, as currently understood, is accepted. The rim can serve as a point of access to the axis and the Center only if it is taken for what it really is, namely the rim. Once the fact that the rim is the periphery is forgotten the center also ceases to possess meaning and becomes inaccessible. Were a true metaphysics, a *scientia sacra*, to be once again a living reality in the West, knowledge gained of man through scientific research could be integrated into a pattern which would also embrace other forms of knowledge ranging from the purely metaphysical to those derived from traditional schools of psychology and cosmology. But in the field of the sciences of man, as in that of the sciences of nature, the great impediment is precisely the monolithic and monopolistic character which modern Western science has displayed since the 17th century. Putting aside the great deal of pseudo-science and simply erroneous theories prevalent in the modern sciences of man such as anthropology and psychology, the elements that are based on careful observation of human behavior or the human psyche under different circumstances could be related, without any logical contradictions, to what traditional schools of psychology such as those found in Sufism, or Yoga or Zen, have also discovered about the human psyche, and especially certain aspects of it of which most modern men remain totally unaware.[9] But this is possible only if the concept of man in his totality as the "universal man" (*al-insân al-kâmil*) of Islamic esotericism and as expounded in traditional metaphysics is accepted, for as mentioned already only the greater can embrace the lesser. But to claim to know the human psyche without the aid of the Spirit (or the Intellect) and to claim a finality for this knowledge as a "truly scientific knowledge" that is independent of any other form of knowledge cannot but result in the impasse with which the modern world is faced today. It can only end in a truncated and incomplete, not to say outright erroneous, "science of man," which is asked to play a role for which it has no competence and which is most often more dangerous than ignorance pure and simple, for there is nothing more dangerous than simple ignorance except an ignorance which has pretensions of being knowledge and wisdom. Scientific research into the nature of man can possess a constructive relationship to the universal and perennial ways of knowing man only if it realizes its own limitations and does not seek to transgress the limits inherent in its approach. It can be legitimate only if it is able to overcome the "totalitarian rationalism" inherent in modern science[10]—although not accepted by many scientists—

9 Unfortunately very few serious studies based on the traditional point of view, which alone matters, have been made as yet of the traditional psychological sciences of the various Oriental traditions, sciences which can be understood only in the light of metaphysical principles and can be practiced only with the aid of the spiritual grace present in a living tradition. See A. K. Coomaraswamy, "On the Indian and Traditional Psychology, or rather Pneumatology," in *Selected Writings of Ananda K. Coomaraswamy*, ed. by R. Lipsey, Princeton, 1977.

10 Frithjof Schuon, *Light on the Ancient Worlds* (World Wisdom, Bloomington, Indiana, 1984) p. 117.

and accept to become what it really is, namely a limited and particular way of knowing things through the observation of their external aspects, of phenomena, and of ratiocination based upon this empirical contact with things; a way that would be acceptable if taken for what it is, because things do also possess a face turned toward the external and the exteriorized.

The answer to the question of what is the worth of scientific research as a source of universal or essential knowledge about man must then be that it is worthless if considered as a source. How can a knowledge which negates the universal order in the metaphysical sense and denies the possibility of essential knowledge serve as a source of knowledge that is essential and universal? Scientific research can become a source of essential knowledge provided it becomes a form of *scientia sacra*, as already mentioned, provided "scientific" is understood in the traditional sense of a knowledge that issues from and leads to the center or the principial order.

There is, however, one way in which scientific research can aid in gaining an awareness of something essential about the present predicament of man, if not of man's eternal nature. This is to make use of the method that science employs in carrying out experiments to study modern scientific and industrial civilization itself. In science whenever an experiment does not succeed, it is discontinued no matter how much effort has been put into the experiment, and an attempt is made to learn from the errors which were responsible for the lack of success of the experiment.

Modern civilization as it has developed in the West since the Renaissance is an experiment[11] that has failed and in fact failed in such an abysmal fashion as to put into doubt the possibility of a future for man to seek other ways. It would be most "unscientific" today to consider this civilization with all the presumptions about the nature of man and the Universe which lie at its basis as anything

---

11 "But, properly, urban industrialism must be regarded as an experiment. And if the scientific spirit has taught us anything of value, it is that honest experiments may well fail. When they do, there must be a radical reconsideration, one which does not flinch even at the prospect of abandoning the project. Surely as of the mid-twentieth century, urban industrialism is proving to be such a failed experiment, bringing in its wake every evil that progress was meant to vanquish." T. Roszak, *Where the Wasteland Ends, Politics and Transcendence in Post-industrial Society*, (Garden City, New York, 1973), p. xxiv of introduction.

other than an experiment that has failed. And in fact "scientific" research if not atrophied by that totalitarian reign of rationalism and empiricism alluded to above should be the easiest way of enabling contemporary man to realize that modern civilization has failed precisely because the premises upon which it has been based were false, because this civilization has been based on a concept of man which excludes what is most essential to the human state.

Paradoxically enough, the awareness of the shortcomings of modern civilization has dawned upon the general Western public— not upon the small intellectual elite who spoke of the crisis facing the modern world as far back as over half a century ago[12]—not because of a sudden realization of man's forgotten nature but because of the rapid decay of the natural environment. It is a symptom of the mentality of modern man that the deep spiritual crisis which has been making the very roots of his soul gradually wither away had to come to his attention through a crisis within the physical environment.

During the past few years so much has been written about the environmental and ecological crisis that there is no need here to emphasize the dimension of the problems involved. The famous study that has emanated from M.I.T., namely *Limits to Growth*, has sought to apply the very methods of modern science to a study of the effects of the application of this science in the future, and the authors of that work as well as many others seriously concerned with the ecological crisis have proposed a change in man's concept of growth, a return to non-material pursuits, a satisfaction with fewer material objects and many other well-meaning changes. But very few have realized that the pollution of the environment is no more than the after-effect of a pollution of the human soul which came into being the moment Western man decided to play the role of the

---

12 Such men as René Guénon in his *Crisis of the Modern World*, trans. by M. Pallis, and R. Nicholson, London, 1962, whose original French edition first appeared in 1927, followed by other traditional authors especially Frithjof Schuon and A. K. Coomaraswamy, have written extensively during the past few decades on the crisis of the West on the basis of the application of perennial metaphysical criteria to the contemporary situation. But their writings were ignored in academic circles for a long time and continue to be so to a large extent even today. The crisis had to appear on the physical level in order to bring the dangerous tendencies of modern civilization before the eyes of modern men.

Divinity upon the surface of the Earth and chose to exclude the transcendent dimension from his life.[13]

In this late hour of human history there are two tragedies we observe, one in the West and the other in the East. In the Occident where the crisis of modern civilization which is after all the product of the West, is most fully felt, since it is related usually to the environmental crisis, solutions are proposed which contain the very factors that led to the crisis in the first place. Men are asked to discipline their passions, to be rational humanists, to be considerate to their neighbors, both human and non-human. But few realize that these injunctions are impossible to carry out as long as there is no spiritual power to curb the infernal and passionate tendencies of the human soul. It is the very humanist conception of man that has dragged him to the infra-human. It is as a result of an ignorance of what man is, of the possibilities of the depths of darkness as well as the heights of illumination that he carries within himself, that such facile solutions are proposed. For millennia religions have taught men to avoid evil and to cultivate virtue. Modern man sought to destroy first the power of religion over his soul and then to question even the meaning of evil and sin. Now many propose as a solution to the environmental crisis a return to traditional virtues, although usually they do not describe them in such terms, because most of them remain secular and propose that the life of men should continue to be divorced from the sacred. It might be said that the environmental crisis, as well as the psychological imbalance of so many men and women, the ugliness of the urban environment and the like, are the results of the attempt of man to live by bread alone, to "kill all the gods" and announce his independence of heaven. But man cannot escape the effect of his actions, which are themselves the fruit of his present state of being. His only hope is to cease to be the rebellious creature he has become, to make peace with both heaven and earth and to submit himself to the Divine. This itself

---

13 We have dealt with this theme extensively in our *The Encounter of Man and Nature, the Spiritual Crisis of Modern Man.*

"What, after all, is the ecological crisis that now captures so much belated attention but the inevitable extroversion of a blighted psyche ? Like inside, like outside. In the eleventh hour, the very physical environment suddenly looms up before us as the outward mirror of our inner condition, for many the first discernible symptom of advanced disease within." T. Roszak, *op. cit.*, p. xvii of introduction.

would be tantamount to ceasing to be modern as this term is usually understood, to a death and a re-birth. That is why this dimension of the problem is rarely considered in general discussions of the environmental crisis. The missing dimension of the ecological debate is the role and nature of man himself and the spiritual transformation he must undergo if he is to solve the crisis he himself has precipitated.

The second tragedy, which is that occurring in the East, or more generally in the non-Western world, is that that world for the most part is repeating the very errors which have led to the failure of urban-industrial society and modern civilization that has produced it, whereas its attitude towards the West should be to view it as a case study to learn from rather than a model to emulate blindly. Of course the politico-economic and military pressures from the industrialized world upon the non-Western world are so great as to make many decisions impossible and many choices well nigh excluded. But there is no excuse for committing certain acts whose negative results are obvious and in having no more reason for undertaking this or that project than the fact that it has been carried out in the West. The earth cannot support the mistakes committed by Western civilization again, and it is most unfortunate that no present-day power on earth has a wide enough perspective as to have the well-being of the whole earth and its inhabitants in mind.

Of these two tragedies, certainly the first overshadows the second, for it is action carried out in the modernized, industrialized world that affects more directly the rest of the globe. For example were the ecological crisis really to be taken seriously by any of the major industrial powers in their economic and technological policies it would have an immeasurable influence upon those who of necessity emulate these powers in such fields. How different would the future of man be if the West were to remember again who man is before the East forgets the knowledge it has preserved over the ages about the real nature of man!

What contemporary man needs, amidst this morass of confusion and disorder of both a mental and physical order which surrounds him, is first and foremost a message as to who man is, but a message that comes from the Center and defines the rim vis-à-vis the Center. This message is still available in a living form in the Eastern traditions and can be resuscitated within the Western tradition. But wherever this message be found, whether in the East or the West, if it issues from the Center, it is always a call for man who lives on the periphery and the rim of the wheel of existence to follow the spokes to the axis or Center which is at once the Origin of himself and of all things. It is a call for man to realize who he is and to become aware of that spark of eternity which he contains within himself. "There is in every man an incorruptible star, a substance called upon to become crystallized in Immortality; it is eternally prefigured in the luminous proximity of the Self. Man disengages this star from its temporal entanglements in truth, in prayer and in virtue, and in them alone."[14] He who has crystallized this star is at peace with both himself and the world. Only in seeking to transcend the world and to become a star in the spiritual firmament is man able to live in harmony with the world and to solve the problems that terrestrial existence by its very nature imposes upon him during this fleeting journey in the temporal which comprises his life on earth.

14 Frithjof Schuon, *Light on the Ancient Worlds* (World Wisdom, Bloomington, Indiana, 1984), p. 117.

6

Christianity and the Religious Thought of C. G. Jung

Philip Sherrard

Two preliminary remarks must preface this chapter. The first concerns the source material on which it is based. Jung had no "religion" in the commonly accepted sense of the word. He did not belong to any branch of the Christian Church, nor did he affiliate himself to any other explicit religious tradition, like Islam or Buddhism. Therefore on the face of it he did not accept any system of doctrine or dogma based on revelation and elaborated by the spiritual interpreters of the tradition in question. On the contrary, he claimed that he was a natural scientist, and that such religious ideas as he had were developed over the course of his life in relation to his empirical experience as a psychologist and the reading he undertook in order to reach an understanding of what he had experienced.

If this is true, then his religious thought was in a continual state of growth and modification. It was fluctuating, rather than stable. What he perceived or believed at certain times might be altered or even reversed by subsequent experience and reading. Therefore one would risk being unfair to Jung if one were to extract concepts and thoughts from the developing body of his work and to say that these represent his religious ideas. One would have to make sure that they were concepts or thoughts he maintained up to the end, and did not reject or modify out of recognition. Consequently, for the purposes of this chapter, it has seemed best to confine attention to his autobiography, *Memories, Dreams, Reflections* (London, 1963), put together during the last years of his life and expressing his ideas at their most mature and most intimate level. This has the additional advantage that this book—at least the English edition—has been supplied with a glossary giving, through extracts from earlier works, explanations of his central psychological concepts and terminology.

The second remark can be put in the form of a question: to what extent are we entitled to speak of "religious thought" at all where Jung is concerned? Religious ideas normally speaking derive from and refer to a world or truths that are regarded as supernatural or metaphysical. They are to do with metaphysical realities. It is not that Jung refused to discuss problems commonly called religious. It is even stated in the introduction to his autobiography that he "explicitly declared his allegiance to Christianity." But, as the introduction continues, he looked at religious questions from "the standpoint of psychology, deliberately setting a bound between it and the theological approach." In fact, this is an understatement, at least where intention is concerned. Jung not only sought to set a bound between psychology and theology. He denied the very basis of theological statement altogether.

This he did from, as it were, both ends. First, he denied the objective existence of those metaphysical or metapsychical realities which theological statements presuppose, and affirmed that there is no truth but purely subjective truth. "We are still a long way from understanding what it signifies," he writes (p. 15), "that nothing has any existence unless some small—and oh, so transitory—consciousness has become aware of it." Then he denied—as a necessary consequence, it might be said, of this initial denial—that there can be any statement or comprehension at all other than the psychological. The passage is worth quoting in full, since it shows how far Jung was willing to go in rejecting the validity of the theological standpoint (at least as theologians themselves understand it), and illustrates the contradictions in which he is involved as a result. "All conceivable statements," he writes (pp. 322-323), "are made by the psyche. . . . The psyche cannot leap beyond itself. It cannot set up any absolute truths, for its own polarity determines the relativity of its statements. . . . In saying this we are not expressing a value judgement, but only pointing out that the limit is very frequently overstepped. . . . In my effort to depict the limitations of the psyche I do not mean to imply that only the psyche exists. It is merely that, so far as perception and cognition are concerned, we cannot see beyond the psyche. . . . All comprehension and all that is comprehended is in itself psychic, and to that extent we are hopelessly cooped up in an exclusively psychic world." There are, in other words, no supra-psychic realities that man can comprehend, and all so-called theological statements that pretend to derive from and refer to such realities are really no more than psychological statements (if that) invested by their authors with a status which in the nature of things they cannot possess.

As is so often the case with those over-anxious to deny a point of view other than their own, Jung in fact is led into a position which contradicts what he wishes to affirm. In saying that "every point of view is necessarily relative" (p. 198), and that "all conceivable statements are made by the psyche," and that "all comprehension and all that is comprehended is in itself psychic," clearly what he wishes to emphasize is that no theological or metaphysical statement has the significance which a theologian or metaphysician would claim for it. It must in the nature of things be subjective, relative, psychic, and refer only to subjective, relative, and psychic realities. We are exclusively doomed to this relative, subjective, psychic world.

Yet if that is the case, Jung's statements themselves are not exempt from these conditions. They too are relative, subjective and psychic. In that case, their categorical appearance is all bluff. Objectively, as enunciations of general truths they can have no significance. To say that "every point of view is necessarily relative" is virtually a meaningless thing to say, since, taken at its face value, then it itself represents but a relative point of view and so cannot apply as a general statement valid for every point of view. For a statement to be valid for every point of view there must be some point of view which is not relative but capable of embracing all points of view.[1]

---

1 *Editor's note:* "Relativism sets out to reduce every element of absoluteness to a relativity, while making a quite illogical exception in favor of this reduction itself. In effect, relativism consists in declaring it to be true that there is no such thing as truth, or in declaring it to be absolutely true that nothing but the relatively true exists; one might just as well say that language does not exist, or write that there is no such thing as writing. In short, every idea is reduced to a relativity of some sort, whether psychological, historical, or social; but the assertion nullifies itself by the fact that it too presents itself as a psychological, historical, or social relativity. The assertion nullifies itself if it is true, and by nullifying itself logically proves thereby that it is false; its initial absurdity lies in the implicit claim to be unique in escaping, as if by enchantment, from a relativity that is declared alone to be possible.

The axiom of relativism is that 'one can never escape from human subjectivity'; if such be the case, then this statement itself possesses no objective value, it falls under

Similarly, if all comprehension and all that is comprehended is in itself psychic, then Jung's statement that "all conceivable statements are made by the psyche" is again virtually meaningless. It has no status at all as a general truth, applicable to all statements, but simply represents Jung's own relative and subjective point of view. It could only have a general validity applicable to all statements on condition that it is true in a non-relative and non-subjective manner. But, Jung says, it is impossible for any statement to be non-relative and non-subjective. Why, then, does Jung make this statement in such categorical terms, as if he were making a pronouncement which applies to all statements? Why, in effect, is he issuing a dogma—one, it is true, designed to undermine the traditional basis of religious dogma, but no less a dogma on that account?

The answer would seem to be fairly clear. Indeed, it is precisely this, that he did wish to undermine the traditional basis of religious dogma, as well as of all theological thought of the traditional kind. He wanted to clear the ground, establish a kind of *tabula rasa* on which to build afresh. So long as the great structure of Christian doctrine and dogma, regarded as sacred and inviolate, stood in the way, his own ideas could make little progress. But if he could show that this structure shared in all the necessary limitations of human thought as he conceived them, and was in fact essentially subjective and relative and psychic, its authority would be shaken. It would be seen to have no greater claims to validity and belief than any other system of thought. Indeed, it might even have fewer claims than other such systems, since these could often point to what is called empirical evidence in their support, whereas many of the dogmatic formulations of Christianity appear to flout such empirical evidence.

Jung's task had therefore a twofold direction. First he had to demonstrate that the claim of theology and dogma to possess a kind

---

its own verdict. It is abundantly evident that man can perfectly well escape from subjectivity, otherwise he would not be man; the proof of this lies in the fact that we are able to conceive both of the subjective as such and of a passing beyond it. For a man who was totally enclosed in his own subjectivity, that subjectivity would not even be conceivable; an animal lives its own subjectivity, but does not conceive it because, unlike man, it does not possess the gift of objectivity." Frithjof Schuon, *Logic and Transcendence,* (Perennial Books, London, 1975), Chapter 1.

of eternal and objective status independent of the judgement and even the consciousness of particular individuals was groundless, and that in the nature of things they could possess no greater or more significant—less relative and subjective—status than any other thought-forms or mental formulations; and this he attempted to do in the way we have shown, by insisting that all statements are made by the psyche and that where our understanding is concerned we are hopelessly cooped up in an exclusively psychic world. And second, he then had to create his own system of thought, and to put it forward not as the truth in a theological sense, but simply as a series of tentative, limited observations based upon his purely pragmatic investigations of the human psyche.

In other words, Jung's system of thought could claim validity not because it was metaphysical, but precisely because it was not metaphysical; and he frequently asserts that unlike the theologians he does not overstep the limit, but bases what he has to say, individual and relative as it is, on solid scientific ground. "My *Answer to Job*," he writes (pp. 206-207), "was meant to be no more than the utterance of a single individual. . . . I was far from wanting to enunciate a metaphysical truth. Yet the theologians tax me with that very thing, because theological thinkers are so used to dealing with eternal truths that they know no other kinds. When the physicist says the atom is of such and such a composition, and then he sketches a model of it, he does not intend to express anything like an eternal truth. But theologians do not understand the natural sciences, particularly, psychological thinking."

This is very disarming, and one might well be taken in by it were it not for the fact that when it comes to the point Jung is quite as capable of making categorical statements lacking all so-called empirical basis as the most dogmatic theologian. Those few already cited could be matched by others occurring throughout the book. Indeed, it is quite clear from a reading of this book that Jung's thought is essentially religious. It may even be said that he regarded himself as the apostle of a new religion, one that should replace for western man the exhausted formulas of Christianity, and one that in this scientific age would stand a far greater chance of acceptance if its own tenets could be presented in the guise of scientific theory, underpinned by solid psychological, observation. Moreover, there seems to be little doubt that Jung regarded his mission as God-given. He was, it may be recalled, the only son of a Protestant pastor, and eight of his uncles were also pastors. Religion, one might say, was in his blood.

Yet it was not the religion of his father, or indeed of Christianity as the Church presented it. This religion was in many ways abhorrent to him. In his youth, he tells us (p. 55) the Church was a place of torment for him, and not until the age of thirty could he confront *Mater Ecclesia* without a sense of oppression (p. 30). Though greatly traveled, he could never go to Rome, and when in old age he at last decided to make the journey he had a fainting fit while buying the tickets and had to turn back (p. 269). But this dislike of the Church and its theology did not mean that he was therefore an atheist. On the contrary—and it is here one can discern to what extent he felt his mission was God-given—he considered that God disliked the Church and its theology just as much as he did, if not more.

When still quite young, he had a terrifying and "sinful" thought. He thought of God sitting high up in the clouds on a golden throne and excreting a large turd which fell on the cathedral of his home-town and shattered it (p. 50). Later, in relation to this experience, he writes (p. 98): "Now I understood the deepest meaning of my earlier experience: God Himself had disavowed theology and the Church founded upon it." Therefore in undermining and denying the basis of traditional Christian theology and in propagating his own gospel in its place, Jung did not feel he was acting in an arbitrary and irresponsible or godless manner. He felt he was carrying out the will of God. He had been entrusted by God with this mission: to make clear to men what God had disavowed and why He had disavowed it; to save God Himself, and man with Him, from the theology and the Church which smothered them, and to proclaim a new religion of life to take the place of a moribund Christianity.

It is this that entitles us to speak of Jung's ideas as religious without doing them any violence. The main lines of his thought, his central concepts and images, constitute what really amounts to a theology and a mythology. Moreover, he did not himself regard this theology and mythology as anti-Christian. As we have seen, he professed his allegiance to Christianity. He thought Christianity of central importance for western man (p. 200). But he considered that it needed to be "seen in a new light, in accordance with the changes wrought by the contemporary spirit." Otherwise, he writes, "it stands apart from the times, and has no effect on man's wholeness" (pp. 200-201). In effect, he thought Christianity had concentrated too much on the ideal, bright, and good side of man's nature, and this at the expense of the non-ideal, dark, and sinful side. When he was a young boy he had a dream of a phallus on a throne in an underground cavern. This had made it difficult for him to accept the conventional image of the Christian Savior. "Lord Jesus never became quite real for me," he writes (p. 27), "never quite acceptable, never quite lovable, for again and again I would think of his underground counterpart, a frightful revelation which had been accorded to me without my seeking it."

According to one interpretation of the dream the phallus represented the dark side of Jesus. Later it was revealed to him as "the breath of life," the "creative impulse" (p. 36). He thought that God, who had disavowed the theology and the Church that had concentrated on the ideal and good side of man, was now wanting "to evoke . . . his darkness and ungodliness" (p. 77). As we had failed, through traditional Christianity, to overcome or escape our anxiety, bad conscience, guilt, compulsion, unconsciousness and instinctuality from the bright, idealistic side, "then perhaps we shall have better luck by approaching the problem from the dark, biological side" (p. 149).

In a way, one might say that Jung regarded it as his task to redeem the Devil. The Devil in Christian thought had come to represent everything that was evil, godless, instinctive, dark in life. All this was regarded as the opposite of God, who was exclusively good, rational, bright. Consequently Christians had concentrated on suppressing all these "diabolic" aspects of themselves and on developing only their good, rational, bright aspects. The result had been the warping and sterility of human life. Now these "diabolic" elements needed to be released and integrated into man's experience of himself.

Moreover, these elements must be seen not as belonging to the Devil as the enemy of God, but as aspects of God's own nature. What traditional Christianity had foisted on to the Devil as a figure diametrically opposed to God and had driven out into the wilderness as a kind of scapegoat, must now be seen to have its source in God. The Devil is also God. God is "the dark author of all created things, who alone was responsible for the sufferings of the world" (p. 97). The chthonic spirit, the spirit indicated in the dream of the underground phallus, is the "other face of God," the "dark side of the God-image" (p. 163). It is God who has created the world and its sins (p. 206), and in failing to recognize this—in failing to recognize that God is the author of evil just as much as of the good—Christianity had promulgated a false idea of God whose acceptance had resulted in the gradual atrophy of man's creative life. Now God was calling for His dark evil side to be recognized and accepted once more, so that the dark evil side of human nature could also be recognized and accepted. Both God and man were seeking to be liberated from imprisonment in the good, ideal, bright, rational side of themselves, so that they could again function in their original wholeness. It was this call for liberation on the part of God and man to which Jung felt compelled to respond. It was this that constituted his religious mission and it was to the realization of this mission that he devoted his creative life and developed his religious thought and mythology.

From an autobiographical and historical point of view, Jung's religious thought may be said to have begun as a reaction against the type of Protestant Christianity represented by his father, and so, by extension, against the extreme rationalism of the nineteenth-century western world. This Christianity seemed to amount to a more or less blind adherence to various articles of faith one never questioned and which effectively cut one off from any real experience either of man or God. "The arch sin of faith, it seemed to me," he wrote (p. 98), "was that it forestalled experience." Together with the passive acceptance of this untested and unlived religious dogma went an elementary moral code based on a clear-cut and equally unquestioned opposition between good and evil, black and white. Living the Christian life seemed to consist solely in maintaining faith in this abstract bundle of Christian precept by turning a deaf ear to everything that assailed it, and in conforming to the prescribed moral code. It was a mixture of mental bigotry and moral will-power. Nor did the rationalism of current nineteenth-century scientific thought, against which men like Jung's father were so stubbornly opposed, seem any more satisfactory. This too seemed solely a device for shutting man off from any living contact with real life. The attempt to dominate everything by the reason seemed but to serve the secret purpose of placing one at a safe distance from real experience and of substituting for psychic reality an apparently secure, artificial, but merely two-dimensional conceptual world in which the reality of life is well covered up by so-called clear concepts. Both the intellectual idealism and ethical dualism of Protestant Christianity and the naïve rationalism of nineteenth-century science seemed to leave out of account and provide no explanation for those realities of which his youthful visions and experiences had made him aware. They seemed to leave out of account and offer no explanation for the whole irrational, dark, primitive, "evil" side of man's nature.

This Jung was able to perceive more objectively when at a later stage in his life he made a journey to North Africa and came into contact with the Arab world. The passage in which he speaks of this, though it includes expressions and ideas deriving from a more fully formulated phase of his thought, deserves to be quoted because it demonstrates what he must have less explicitly realized when at the outset of his intellectual development he reacted against the religion of his childhood and the rationalism of modern western man. "The emotional nature of these unreflecting people [the Arabs]," he writes (pp. 230-231), "who are so much closer to life than we are exerts a strong suggestive influence upon those historical layers in ourselves which we have just overcome and left behind, or which we think we have overcome. It is like the paradise of childhood from which we imagine we have emerged, but which at the slightest provocation imposes fresh defeats upon us. . . . The sight of a child or a primitive will arouse certain longings in adult, civilized persons—longings which relate to the unfulfilled desires and needs of those parts of the personality which have been blotted out of the total picture in favor of the adopted persona. . . . The predominantly rationalistic European finds much that is human alien to him, and he prides himself on this without realizing that his rationality is won at the expense of his vitality, and that the primitive part of his personality is consequently condemned to a more or less underground existence."

Since the lifeless abstractions of the Christian faith, though supposedly referring to supernatural realities, and the two-dimensional conceptual world of the rationalists were both creations or at least appurtenances of man's everyday consciousness, and belonged to what he consciously believed or thought, Jung found it convenient to give that primitive part of the personality which modern man had condemned to a more or less underground existence an opposite label and to call it the unconscious. This concept of the

119

unconscious, later elaborated into the "collective unconscious," is crucial to Jung's whole system of thought, so it is important to try to get clear what he meant by it.

This is not an easy task, since in spite of its crucial position in Jung's thought it nevertheless remains a somewhat vague concept. To begin with, Jung seems to have thought of the unconscious as a kind of repository of all those psychic elements and drives which have either not entered man's conscious world, or been driven out of it, suppressed, because of his inability or unwillingness to admit them on the conscious level. He first became graphically aware of it in a dream. In this dream, Jung found himself in a two-story house. He was on the upper floor. He first descended to the ground floor, and then to the cellar, and finally down into a cave cut out in the rock beneath the house, where there were two human skulls lying among scattered bones and broken pottery (p. 155). He interpreted the dream as a kind of structural diagram of the human psyche. The upper floor where he first found himself represented the consciousness; the ground floor stood for the first level of the unconscious, while the cave itself was the world of the primitive man in every human being. This primitive and deepest part of man's unconscious psyche borders, he writes, on the animal soul, just as the caves of prehistoric times were usually inhabited by animals before men laid claim to them (p. 156).

According to this interpretation, the dream appeared to postulate "something of an altogether *impersonal* nature" underlying the psyche (p. 157). This, Jung says, was his first inkling "of a collective a priori beneath the personal psyche." He took it to be "the traces of earlier modes of functioning." Later, "with increasing experience and on the basis of more reliable knowledge," he recognized these earlier modes of functioning "as forms of instinct, that is, as archetypes" (p. 157). These archetypes and forms of instinct ("archetype" and "instinct" are synonyms in Jung's terminology, and their sense must on no account be confused with that of "archetype" in the traditional Platonic and Christian meaning of the word) constitute the unconscious. He calls this unconscious "collective" because, "unlike the personal unconscious (represented by the ground floor in the dream), it is not made up of individual and more or less unique contents but of those which are universal and of regular occurrence. . . . The deeper 'layers' of the psyche lose their individual uniqueness" (p. 357, note on the *Unconscious*). The collective unconscious is common to all people (pp. 136-137) and it consists of "archaic psychic components which have entered the individual psyche without any direct line of tradition" (p. 35).

This schematic representation of the psyche in the form of a house gives what one might call its vertical cross-section. But this vertical progression from lower to higher, from the cave to the upper story, has its corresponding horizontal extension in man's actual historical evolution. Although Jung had become aware of limitations in the rationalism of nineteenth-century scientific thought, he did not on that account reject its theories, or at least not all of them; and one of these theories which he accepted totally and interwove with his own thought so intimately that one can say that they stand or fall together, was the Darwinian theory of man's evolution. This theory he married to his own conception of the human psyche, and particularly of the unconscious. That is to say, he thought that the various layers of the human psyche had their counterparts in the various phases of man's evolution through the centuries on earth.

The conscious aspect of the psyche represented man's present phase of evolution; those aspects of the psyche which in the course of his evolution western man condemned to a more or less underground existence correspond to those historical layers in himself that he has overcome and left behind, but that still remain buried within him. Thus it is that the deepest level of the collective unconscious, the deepest part of man's nature, "borders on the life of the animal soul" (p. 156). This correlation of the psyche with man's evolutionary progress led him, inevitably, to reject Christian ideas of man's creation and consequently of the structure of the human psyche, and to substitute his own ideas. "If the unconscious is anything at all," he writes (p. 320), "it must consist of earlier evolutionary stages of our conscious psyche. The assumption that man in his whole glory was created on the sixth day of Creation, without any preliminary stages, is after all somewhat too simple and archaic to satisfy us nowadays. There is pretty general agreement on that score. . . . Just as the body has an anatomical pre-history of millions of years, so also does the psychic stream. And just as the human body represents in each of its parts the result of this evolution, and everywhere still shows traces of its earlier stages—so the same may be said of the psyche. Consciousness began its evolution from an animal-like state. . . ."

The Christian idea of man's psyche—that man's consciousness has its roots in the Divine and that only as a consequence of his degradation and immersion in earthly and animal existence has it become obscured—would seem here to be turned on its head: human consciousness began in the dark subhuman world of the animals and plants and over the centuries has been gradually emerging into the light of complete evolution. Perhaps nowhere does Jung's thought appear to be more non-Christian, not to say anti-Christian, than in relation to this crucial concept, the idea of the collective unconscious.

Man's psyche, then, is made up of its conscious and its unconscious components; and although fully-evolved consciousness is the goal towards which man's life is ultimately directed, only too often his actual state of consciousness, even in this latter-day phase of his evolution, is pathetically meager. Largely through an over-development of his reason (though Jung more frequently calls the reason the intellect, not recognizing any distinction between these two faculties[2]), and his refusal to admit into consciousness anything which is not rational or capable of being rationalized, he has driven underground, suppressed, locked up in the unconscious all those primitive, irrational, instinctive contents of the psyche on which his vitality and creativeness depend. To such an extent has he done this that what one may take to be the proper relationship between consciousness and the unconscious in fully-evolved man has now been reversed; and far from man being truly aware of what he is, what he thinks he is bears little or no relationship to his total being.

In fact, his true life now is not his conscious life at all, but his unconscious life. "Our unconscious existence," he writes (pp. 299-300), "is the real one and our conscious world is a kind of illusion, an apparent reality constructed for a specific purpose, like a dream which seems a reality as long as we are in it." The tenacity with which nonetheless modern western man clings to this conscious world—this illusion—at the expense of the unconscious world results not only in reducing his existence to a kind of shadow-play but also in chronic psychic dislocation. It is in fact the

---

2 *Editor's note:* For a clarification of this distinction see Chapter 9 in Philip Sherrard's book *Christianity: Lineaments of a Sacred Tradition.* (Holy Cross Orthodox Press, Brookline, Massachusetts, 1998.)

extreme resistance which the consciousness of modern western man offers to the unconscious contents of the psyche that produces the vast range of psychic disorders, both individual and collective, that characterize our time. Modern man has identified himself with what is at best but a superficial aspect of himself, and his real self lies buried within him, in the obscure substrata of the unconscious. It follows from this that if modern man is to recover his psychic health and realize what Jung calls his wholeness or complete form he must once again allow his submerged, suppressed, unconscious existence to enter his conscious world.

To this process whereby man bit by bit releases the submerged contents of his unconscious into consciousness and so achieves his wholeness of being Jung applies the term *individuation.* Individuation, the glossary states (p. 352), "means becoming a single, homogeneous being, and, insofar as individuality embraces our innermost, last, and incomparable uniqueness, it also implies becoming one's own self. We could therefore translate individuation as 'coming to selfhood' or 'self-realization'." It must not on any account be confused with the coming of the ego into consciousness, which results simply in ego-centeredness and auto-eroticism. The self, that has to be realized through individuation, comprises infinitely more than the ego. The self embraces not only the conscious, but also the unconscious psyche; and it does not simply embrace them, it is the center of this totality, just as the ego is the center of the conscious mind.

The self is the wholeness of the personality (p. 187), that which we are, the "principle and archetype of orientation and meaning" (p. 190). It is realized through a process of self-knowledge by means of which "we approach the fundamental stratum or core of human nature where the instincts dwell. . . . This core is the unconscious and its contents" (p. 305). Through self-knowledge the psyche is transformed by changing the relationship between the ego, or human consciousness in the ordinary restricted sense, and the contents of the unconscious. What was before hidden and forced underground is now brought out into the open and liberated.

In this psychic transformation what Jung calls the *anima* (or, in woman, the *animus)* plays a vital part. The *anima* is in a sense the transforming instrument, the go-between operating between the conscious and the unconscious world. She is a kind of psychopompos, establishing the relationship with the unconscious.

Related to the unconscious in this way, she has, like the unconscious, a strongly historical character. "As the personification of the unconscious she goes back into prehistory, and embodies the contents of the past. She provides the individual with those elements that he ought to know about his prehistory. To the individual, the *anima* is all life that has been in the past and is still alive in him" (p. 267). She functions thus as a bridge or a door, leading into the unconscious. There, in the unconscious, she produces a mysterious animation, and gives visible form to ancestral traces, the collective contents (p. 183). Like a medium, she gives these contents a chance to manifest themselves.

This they do in terms of images and myths. Images and myths are not human inventions. They are the spontaneous forms in which the unconscious reveals itself. Projected by the unconscious into the *anima* they can be grasped by the consciousness. Received in the consciousness, and there interpreted, they provide the means through which the contents of the unconscious are released into the consciousness. It is through myth and symbol that individuation is achieved. Through them, man can begin to live all those phases of his evolutionary past that are still present within him. Through them too he comes into contact with his primitive instinctive life, with those pre-existent dynamic factors which ultimately govern the ethical decisions of his consciousness (p. 305). Consciousness and unconscious are thus brought into relationship and harmony. Man reaches the goal of his psychic development, represented by the self. He achieves his wholeness.

These notions of the unconscious, the *anima,* and the process of individuation could be derived, Jung claimed, from empirical observation of his own and other people's psychic activity, although, as he admits (p. 192), psychology is subject far more than any other science to the personal bias of the observer. In fact, he seems to have arrived at them after a lengthy and dramatic confrontation with his own unconscious (see chapter 6 of Jung's *Memories, Dreams, Reflections*), a confrontation which lasted over some eight years (1912-1920) and on which he embarked after his split with Freud. Hence it could be maintained that these notions are purely scientific notions and do not enter into the sphere of religious ideas at all. But not only are they so embedded in his religious ideas that it is impossible to speak of the latter without including them; they also—and this is more important—presuppose the acceptance of certain ideas which in this context one can only call religious.

This is particularly true where the process of individuation is concerned. The process of individuation, or psychic transformation, through which the contents of the unconscious are released into the consciousness depends entirely on the understanding and interpretation of the myths and symbols in which these contents have revealed themselves to the consciousness. Their meaning and significance must be known and recognized. If they are not understood and interpreted there is danger of miscarriage and the whole process of psychic development is in danger of failing or at least of being arrested.

This means that before one can successfully complete the process of psychic transformation one must already be in possession of certain a priori principles of understanding and interpretation in the light of which one can give significance to the myths and images which the *anima* bears from the unconscious to the consciousness. Without these principles, one is simply working in the dark. There is no objective pattern of meaning, nothing according to which one can read the signs in which the unconscious is urgently seeking to transmit its messages. The process of *individuation* therefore presupposes the acceptance of certain ideas, certain principles of understanding, which cannot themselves be derived from empirical observation but which must be applied as it were *ab extra* to the psychological process that is being observed. This is an inescapable condition of individuation. Its implications are considerable. They lead directly into the sphere of what in this context can only be called religious ideas.

Jung already recognized this, though perhaps not clearly, at quite an early stage in his career. His interest in mythology started before his own personal confrontation with the unconscious. It started, he writes (p. 158), in 1909, when he "read . . . through a mountain of mythological material, then through the Gnostic writers . . ."; and it is evident that already the basic principles according to which he began to interpret the significance of the fantasies he experienced during his long personal confrontation with the unconscious were derived from his reading at this time. But after his personal confrontation, the need to find an objective standard of reference, a structure of a priori ideas allowing him to interpret the significance of the myths and images thrown up by the unconscious, became far more pressing. To understand the fantasies arising from this confrontation, he had, he writes (p. 192), "to find evidence for the historical prefiguration of my inner experiences." He had to discover where the premises underlying his experiences had already occurred in history. Unless he could do this, he writes (p. 192), he would never have been able to substantiate his ideas, since a psychologist depends in the highest degree upon historical and literary parallels if he wishes to exclude at least the crudest errors in judgement.

In this crucial quest, Jung once more turned to the Gnostics. Between 1918 and 1926 he again "seriously studied the Gnostic writers, for they too had been confronted with the primal world of the unconscious" (p. 192). Whether it was because they provided a confirmation and elucidation of Jung's own personal experiences and intuitions, or whether it was because they consciously or unconsciously began increasingly to condition those experiences and intuitions themselves, there is no doubt that it was in the central religious ideas of the Gnostic writers that he found that objective pattern of meaning, that framework of a priori principles of understanding and interpretation, according to which he evaluated the significance of the myths and images not only of his own and his patients' collective unconscious but also of the various religious systems in which they had in the past been enshrined. They were the historical configuration for which he sought; and having discovered them and accepted them, he applied them with faithful conformity in his interpretation both of the dreams and fantasies he encountered in the course of his professional work and of such Christian and Biblical themes as the dogma of the Trinity or the story of Job. He also applied them in his interpretation of alchemical symbolism.

Jung's interest in alchemical symbolism seems to have arisen directly out of his reading of the Gnostics. Though the Gnostics provided him with his basic theological notions, he felt they were too remote in time to link up immediately with the psychological questions of today (p. 192). Accepting as he did the Darwinian hypothesis of evolution and applying it to the life of the psyche through pre-historical and historical times, he had to find a corresponding evolutionary progress in the symbolic patterns in which the various stages of the emerging subterranean life of the unconscious had been reflected during the last two thousand years or so—

between, that is, the time when the Gnostics had confronted the unconscious and the years in which he had confronted it. At first he failed to discern this progress, and it seemed that the tradition which might have connected the Gnostics with the present had been severed. Finally however he thought he discovered it, in the works of the alchemists. Alchemy, he claimed (though his claim is based on little other than his desire to establish the connection) represented the historical link with Gnosticism. "Grounded in the natural philosophy of the Middle Ages, alchemy formed the bridge on the one hand into the past, to Gnosticism, and on the other into the future, to the modern psychology of the unconscious" (p. 193). In alchemy, Jung writes (p. 196), "I had stumbled upon the historical counterpart of my psychology of the unconscious. The possibility of a comparison with alchemy, and the uninterrupted intellectual chain back to Gnosticism, gave substance to my psychology"; and it was through the understanding of alchemical symbolism that he arrived at the central concept of his psychology: the process of individuation (p. 200).

The chain was—or appeared to be—complete; and to demonstrate it Jung wrote his *Psychology and Alchemy* and his monumental *Mysterium Coniunctionis*. But this seemingly unbroken chain and its demonstration in Jung's works should not blind one to the fact that if it was the understanding of alchemical symbolism that led Jung to the central concept of his psychology, and so appeared to give it an objective authenticity, yet it was on the religious theories of the Gnostics that he based his understanding of alchemical symbolism itself. It was these theories that gave him his doctrinal premises. They are the ultimate key to his psychology and to the significance which he attributed not only to alchemy but also to human life in general.

We have already glanced briefly at certain of these theories, though without referring to their Gnostic antecedents. Their point of departure in Jung's case would seem to lie in his often repeated conviction that God is the author of evil and suffering as well as of good. Put in Biblical terms, God in His omniscience created everything so that the original parents of mankind—Adam and Eve—would have to sin. It was God's intention that they should sin (p. 49). This leads inescapably to the conclusion that in the final analysis God is responsible for the sins of the world (p. 206). The conventional Christian idea of God, therefore, as essentially and exclusively good, and as the author only of what is good, must be modified. God not only includes goodness in His nature; He also includes evil. He is a *complexio boni et mali.*

In fact, not only good and evil are bound together in God; all opposites are bound together in Him. He is female as well as male. This Gnostic idea of a God who is the union of all opposites in a total complex form is at the basis of Jung's religious thought. In its light he refashioned the traditional Jewish and Christian idea of God. The God of the Old Testament is now shown to be a half-Satanic demiurge. The Christian Trinity is enlarged to a Quaternity, with the Devil as a holy fourth. In the account of Job's sufferings "we have a picture of God's tragic contradictoriness" (p. 206). Job is a prefiguration of Christ. He, like Christ, though to a lesser degree, is the suffering servant of God. Both had to suffer because of the sins of the world. It is God who is responsible for these sins. Because of His "guilt" in creating a world full of evil, God must perform an act of total expiation. This He does in subjecting Himself to ritual killing in the Crucifixion of Christ. It is this act of Christ that individual Christians seek to imitate in their lives. In this manner they help God atone.

From what has been said it is clear that in spite of his protests to the contrary, Jung oversteps the boundaries between psychology and theology all along the line. His claim to reject the metaphysical and to restrain himself to the psychological is merely a device for attempting to make the psychical the only legitimate metaphysics. His psychology is virtually a new religion. It is true that this religion, unlike the frankly metaphysical religions, is not concerned with the relationship between the human soul and the supra-psychic transcendent Reality that nevertheless acts intimately upon and within the human soul. It is concerned with the relationship between the consciousness and those psychic events which do not depend upon consciousness but take place on the other side of it in the darkness of the psychical hinterland. It is a religion of pure psychic immanence. It is in confronting the soul's own immanent contents that man encounters the Divine. And this soul is simply the soul as it is, the "materialized" soul, not the soul detached and purified from "earthly" influences and contradictions.

There is no question of any birth of the supra-psychical Spirit in the soul. There is only the realization of the psychic self. Again it is true that Jung called this self the *imago Dei* in man. But in its realization man calls back the projection of his self onto a God outside or beyond him. It is not that this God is explicitly denied. It is simply that He is not any longer regarded, as He is by the metaphysical religions, as the initiator and perfecter through His deifying power of that process of psychic transformation which is completed only when God has *taken the place of the psychic self.* For the practical purpose of the realization of the Jungian self, God is unnecessary. One need neither pray nor have faith nor require the assistance of grace. It is the self alone that is regarded as the unifying, "deifying" power, the regulator and balancer and harmonizer of the conflicting forces in man. The self, in other words, assumes in man the status and function of the Gnostic God. It is the incarnation of this God—a god who unites male and female, good and evil, in a wholeness in which all opposites are integrated. In this way man becomes his own deity. He is the final form of the Gnostic *complexio*—Christ and Satan in one—now destined to appear on the earth as the identity of God and man.

Set in its historical context, Jung's psychology can be seen as a much-needed protest against the simplifications of scientific rationalism. It is a plea for man to face the realities of his own inner world, to take his own path in the fulfillment of his personal created destiny, and not to barricade himself, as is so often the case, behind an abstract structure of religious or metaphysical principles whose only real function is to prevent him from ever realizing who or what he is, to prevent him from ever developing the potentialities of his own unique being. "Anyone who takes the sure road is as good as dead," he writes (p. 27), and against this death he proclaimed "the risk of inner experience, the adventure of the spirit" (p. 140). He sought to affirm what the mechanistic attitude of the modern western mind ignored or denied—man's deep affinities with the natural world, the world of animals and plants, the beauty of earth and sky. He wished to see the spirit of life recognized in everything, not only in man but also in inorganic matter, in metal and stone; and he held that the phenomena of the natural world were expressions of the same energy—psychic energy, as he called it—as that which underlay the various phenomena of the human soul (p. 201). In his study of myths and symbols, he asserted against those who saw in these nothing but futile speculation or childish fantasies their prime significance as the spontaneous irreplaceable language of the human soul. "No science will ever replace myth," he wrote (p. 313), "for it is not that 'God' is a myth, but that myth is the revelation of a divine life in man." But together with, or in spite of, all this, he accepted that hypothesis of scientific rationalism which perhaps more than any other is inimical and stultifying to man's inner growth—namely, the Darwinian hypothesis of evolution. And he also accepted a "metaphysic" which by affirming the idea of a purely immanent deity led inescapably to an idea far more dangerous to human life than atheism itself—the idea of man as a naturally deified or divine creature. Man cannot take the place of God, since man's being can never attain to the essence of God. When nevertheless man tries to take the place of God, he steps over into the sphere not of God, but of the infernal powers of his own soul.

In his autobiography, Jung recounts (pp. 207-208) a dream he had quite late on in his life. In this dream, Jung and his father enter a house. They come into a large hall which was the exact replica of the council hall of Sultan Akbar at Fatehpur Sikri. Only from its center a steep flight of stairs ascended to a spot high up on the wall. At the top of the stairs was a small door, and Jung's father said to Jung: "Now I will lead you into the highest presence." Then he knelt down and touched his forehead to the floor. Jung imitated him, likewise kneeling, but he could not bring his forehead quite down to the floor. Meditating on this dream afterwards, Jung declares that this failure to put his forehead to the floor in the dream "discloses a thought and a premonition that have long been present in humanity: the idea of the creature that surpasses its creator by a small but decisive factor."

In ancient Iranian literature there is a story of a primeval king, *Yima*, whom the highest god, *Ahura Mazdah*, sets over the world he has made, to protect it and nourish it. This *Yima* does. In response to his sacrifices the gods free man and cattle from death, and water and trees from drought. They give *Yima* command over all lands, and also over all the demons, so he can free *Ahura Mazdah*'s creatures from evil. But in the course of time the world falls into materiality and *Ahura Mazdah* says he will send a greater winter over the earth so that no creature can live on it. *Yima* is told to make a fold, a kind of fortress, and there to gather the seed of all living things. This *Yima* does as well.

Then, however, *Yima* begins to extol himself. He begins to think that all that has happened, all the great benefits that have come to the world and its creatures, has happened and have come as a result of what he is and what he has done. He begins to see himself as the real lord of creation, and to ascribe his mastery over all the powers of nature and his own being to himself, and to boast that he is the author of life and immortality. At precisely this moment his royal glory leaves him and he falls into the grip of demons who drive him over the face of the earth and eventually destroy him. When *Yima* regards himself, and what he is in his own created existence, as self-sufficient, and so feels that he is relieved of all need to look for true being beyond himself, he relegates his god and creator, *Ahura Mazdah,* to the realm of the unnecessary. He virtually proclaims himself his own creator. It is at this point of self-assertion that he falls into the power of the demons that eventually destroy him.[3]

It is difficult to avoid the conclusion that Jung's own thought, culminating in the idea of the creature that surpasses its creator by a small but decisive factor, attains an identical point of self-assertion, with all the disastrous consequences this has for the integrity of human life.

---

3 *Editor's note:* "People generally see in Jungism, as compared with Freudism, a step towards reconciliation with the traditional spiritualities, but this is in no wise the case. From this point of view, the only difference is that, whereas Freud boasted of being an irreconcilable enemy of religion, Jung sympathizes with it while emptying it of its contents, which he replaces by collective psychism, that is to say by something infra-intellectual and therefore anti-spiritual. In this there is an immense danger for the ancient spiritualities, whose representatives, especially in the East, are too often lacking in critical sense with regard to the modern spirit, and this by reason of a complex of 'rehabilitation'; also it is not with much surprise, though with grave disquiet, that one has come across echoes of this kind from Japan, where the psychoanalyst's 'equilibrium' has been compared to the satori of Zen; and there is little doubt that it would be easy to meet with similar confusions in India and elsewhere. Be that as it may, the confusions in question are greatly favored by the almost universal refusal of people to see the devil and to call him by his name, in other words, by a kind of tacit convention compounded of optimism to order, tolerance that in reality hates truth, and compulsory alignment with scientism and official taste, without forgetting 'culture,' which swallows everything and commits one to nothing, except complicity in its neutralism; to which must be added a no less universal and quasi-official contempt for whatever is, we will not say intellectualist, but truly intellectual, and therefore tainted, in people's minds, with dogmatism, scholasticism, fanaticism, and prejudice. All this goes hand in hand with the psychologism of our time and is in large measure its result." From a letter of Frithjof Schuon to Titus Burckhardt (published in *The Essential Titus Burckhardt,* ed. William Stoddart, [World Wisdom, Bloomington, 2003], Chapter 2).

7

On Earth as It Is in Heaven

James S. Cutsinger

"It is the Spirit that quickeneth; the flesh profiteth nothing."

John 6:63

"The forces do not work upward from below,
but downward from above."

Hermes

It is often supposed that emanation, creation, and evolution cannot be reconciled. Either the world proceeds of necessity out of God Himself, or God freely chooses to make the world from nothing, or the world contains its cause within itself; its source is either transpersonal, personal, or infrapersonal. The following considerations[1] are designed to show that a reconciliation of these points of view is permissible if it is understood that the cosmogonies in question reflect various angles of approach. This is not to say, however, that each of these perspectives is equally adequate. If the "infrapersonal" explanations of physical science are to have any worth, their dependence upon the "personal" account of theology must be fully acknowledged, even as the personal must in turn admit the priority of the "transpersonal" truths of pure metaphysics. Whatever value there may be in the idea of evolution becomes apparent only in light of creation, and then only to the measure that the doctrine of creation is itself illumined by emanation.

The primary aim of this article is to present an account of an evolving world *per ascensum* fully consistent with the principles presupposed *per descensum* by metaphysics and theology. Not all

1 This is a revised version of an article first published by *Dialogue and Alliance*, Vol. 4, No. 4, Winter 1990-91, and reprinted in *Sacred Web: A Journal of Tradition and Modernity*, Vol. 1, July 1998. Permission to reprint is gratefully acknowledged.

Platonists have been Christians, of course, nor have all Christians been Platonists, but there is such a thing as Christian Platonism. This should be sufficient to show that doctrinal positions which include creation within the perspective of emanation, whether widely persuasive or not, are already a matter of record, and that it is therefore not illegitimate to speak of the metaphysician and the-ologian as sharing certain common "principles". The question remains, however, whether these principles might serve to inform an evolutionary understanding of the physical world. In its trans-formist or Darwinian versions, the theory of biological evolution has been for good reason thoroughly rejected, not only by orthodox theologians, but more importantly by those among our contempo-raries who have most forcefully propounded an integral meta-physics, and who have done the most to promote in that light an esoteric interpretation of the doctrine of creation *ex nihilo*.[2] It is therefore essential that the reasons for this rejection be carefully considered before presenting a truly principial theory of evolution, one which is fully consistent with the "degrees of Reality" and which acknowledges the prerogatives of a *scientia sacra*. We aim in this way to avoid the absurdities which usually accompany modern scientific cosmogonies while at the same time "saving the appearances" in the sequence of certain natural forms.

But before proceeding to an evolutionary explanation of things, we would do well to be reminded first of the distinguishing features of a fully metaphysical theology and of the account of the cosmos provided by an "emanational creation".

*

Either the world proceeds of necessity out of God Himself, it was said, or God freely chooses to make that world from nothing: *Tertium non datur.* Or so at least it has seemed to those whose eager-ness to protect the freedom, and hence the sovereignty, of God has caused them to neglect not only the meaning of "nothing" in the

---

2 We are referring to the authors of the "traditionalist" or "perennialist" school, especially Frithjof Schuon, whose influence will be noticed throughout these obser-vations. For an introduction to Schuon's work, see our *Advice to the Serious Seeker: Meditations on the Teaching of Frithjof Schuon* (Albany, New York: State University of New York Press, 1997).

crucial phrase *ex nihilo*, but also the absoluteness and infinity of God Himself. "Nothing" must here signify either one of two things: that the world is fashioned from no "thing", from no already determinate entity or entities, and of course it is not, for to speak of a cosmos is to speak of limitation, and the origin of limited "things" must of necessity be unlimited; or that the world is made *ex nihilo praeter Deum*, from nothing other than God, for the Source of the being by which existing things are is quite evidently not an absolute vacuum, nor is such a conception even possible: *Ex nihilo nihil fit.* But if when one says that the world is created from nothing, it is accordingly meant that the source of the world is supraformal and Divine, then the expression is clearly not a negation, but simply the transposition into theological terms of emanation. The conception of "nothing" is far from a luxury, of course. It serves to remind us of the impotence and contingency of creatures, of their distance from the God whence they come, and of the "presence of absence" in their make-up; and it can provide in this way a useful corrective for metaphysical expositions in which the discontinuous character of emanation has not been sufficiently stressed, and which therefore risk the error of pantheism. But this error is by no means an inevitable feature of every transpersonal cosmogony.

Nor apart from the transpersonal and metaphysical is the personal or creationist perspective free from certain risks of its own, for it tends not to consider with due care the implications, and even more so the limits, of the Divine Qualities or Names, of which freedom is only one. Theologians who mistrust metaphysicians for fear that the necessity of manifestation, unlike the gift of creation, will compromise the absoluteness of God, and thus His freedom from determination, seem not to have recognized that absoluteness imposes its own limits, not *extrinsic* to be sure, but *intrinsic* and proceeding from the essence of the Divine Reality Itself. To call God omnipotent is not to say that He can do anything, lest the goodness of sovereignty be marred by a purely arbitrary deployment. It is to say instead that He cannot be constrained "from the outside in". But God may, and indeed must, be constrained by His essence, which He cannot negate, any more than a man can lift himself off the ground. Obviously God cannot lie, because He is the truth, and to lie would be to undo the truth of His Word—that Word which He not only speaks, but which, "being of one substance with the Father" (Nicene Creed), He himself is. Though perhaps somewhat less obvious, it is equally certain that God cannot but manifest Himself, whether we call the result of this Self-expression an emanation or a creation. Again the necessity flows from the essence. If absolute, God is "loosened" or freed from all limits; He is unbounded and infinite. But having no bounds, nothing being able to contain or enclose Him, God cannot but pass "outside" of Himself and into the nothing from which, as it were by displacement, He makes His creatures come into being. "None is good but God" (Luke 18:19), and it is the very nature of the only Good to "communicate Itself" (St Augustine).

Although we are using a "personal" language in order to show that the emanational perspective need not be opposed to theological considerations of God and "His" Essence or "His" creatures, it should be clear that if the theologian took Divine Sovereignty with complete seriousness, as he claims to be doing in rejecting the idea of cosmic necessity, he would be obliged to admit the priority of the transpersonal, and hence the legitimacy (to say the least) of metaphysics. For a Reality truly sovereign and truly free is not precisely a being at all, let alone a person, for these, like all categories, must impose their own determinations. Freedom and necessity are thus seen to be one, and their apparent opposition may be resolved. The world is able to proceed of necessity from out of the Divine so as to serve as a manifestation of its Source precisely because God is free from all the constraints that might otherwise condition His nature and interfere with His being true to Himself, or with His wish to express Himself through his Word. The inevitability of the world is in this sense intended.[3]

A few remarks should perhaps be added concerning the "tense" of emanation and creation. The theological account of the origination of things is for various reasons less likely than the metaphysical

3 This synopsis is meant to reflect a perspective much more thoroughly developed by Frithjof Schuon in "Dialogue Between Hellenists and Christians", *Light on the Ancient Worlds*, trans. Lord Northbourne (Bloomington, Indiana: World Wisdom Books, 1984); "Creation as a Divine Quality", *Survey of Metaphysics and Esoterism* (Bloomington, Indiana: World Wisdom Books, 2000); "*Ex Nihilo, In Deo*", *The Play of Masks* (Bloomington, Indiana: World Wisdom Books, 1992); and "Theological and Metaphysical Ambiguity of the Word *Ex*", *The Eye of the Heart: Metaphysics, Cosmology, Spiritual Life* (Bloomington, Indiana: World Wisdom Books, 1997).

to admit that God is "always" making the world, even though the eternality of creatures *in divinis* is necessarily implied by their inevitability, just as their inevitability is by Divine power, for "it is alike impious and absurd to suppose that there was a time when Goodness did not do good and omnipotence did not exercise Its power" (Origen). The theologian often objects, not in this case that God has suffered demotion, being constrained to act in a way He would otherwise not, but that the creation has benefited from an unjustified, and blasphemous, promotion, being accounted "as old as" God. But this is to neglect the fact that not all ontological dependence exhibits itself in chronological succession, as the *filiation* and *procession* of the second two Persons of the Holy Trinity should be enough to establish. Whenever there is a mind there is thought, and whenever a sun there is light, and yet the relationship remains in both instances causal and asymmetrical. And so, despite its eternality, "the universe was created by God, and there is no substance which has not received its existence from Him" (Origen). Whatever their disputes as to whether the world has "always" existed, the metaphysician and theologian are agreed that "as long as" there has been a cosmos, God has "always" been responsible for it. The universe is never a *fait accompli*. It is in each instant, or better between every instant, being brought forth—*by* God, the theologian will say, *from* God, in metaphysical terms; or as above, and according to a certain fusion of perspectives, *by God from nothing other than God*. The personal and transpersonal explanations are thus alike in insisting that the contingency of all that is other than the Supreme Reality is such that the cosmos cannot stand on its own, however briefly, but requires, in order to be, the continuous infusion of uncreated power. Whether one thinks of the world as an emanation or a creation, the tense of the process is present: it *is* streaming forth; it *is* being made.

*

The world as a whole, an ordered whole, and not successively or piece by piece, is a continuous *production of* God or (one might say) *eduction from* God, according to one's angle of vision—with the "process" in either case being a movement "from above to below". Altogether different of course is the explanation of the world provided by the transformist or Darwinian account of evolution, which rests upon a total negation of God, the true Source of all creatures, and which purports to account for the variety of natural forms by means of strictly natural processes, "upward from below". Six reasons for rejecting the theory may be adduced.[4] If these reasons are only briefly considered here, it is because the chief point of this article is not to examine in full all the many problems with the transformist theory, but to present a very different view of evolution, one which takes these criticisms seriously, which builds upon them, and which conforms to the metaphysical perspective they in part reflect.

In the first place, the explanation of order by "natural selection", as even its more honest proponents admit, is no more than a *theory*, and one moreover which can never be tested, let alone proved.[5] Like all scientific theories, it is inevitably tentative and probabilistic, being not even so solid or certain as the empirical data it is meant to explain, themselves of course always subject, if not to doubt or denial, then to continual reinterpretation. Having therefore only provisional force, because an origin strictly inductive, the Darwinian doctrine simply cannot compete at the same level of truth with either metaphysics or theology. Of these, the former proceeds from *noesis* or intellection, which involves the direct appre-

---

4 In presenting these reasons, we have been assisted especially by E. F. Schumacher, *A Guide for the Perplexed* (New York: Harper and Row, 1977), Ch. 9, Sect. II; Martin Lings, *Ancient Beliefs and Modern Superstitions* (London: Unwin, 1980), Ch. 1; Huston Smith, *Forgotten Truth: The Primordial Tradition* (New York: Harper and Row, 1976), Ch. 6; Seyyed Hossein Nasr, *Knowledge and the Sacred* (New York: Crossroad, 1981), Ch. 7; and Titus Burckhardt, "Cosmology and Modern Science", *The Sword of Gnosis: Metaphysics, Cosmology, Tradition, Symbolism*, ed. Jacob Needleman (London: Routledge and Kegan Paul, 1986). The specific formulations, however, and accordingly all the infelicities, are our own.

5 Of course the theory has been accepted as indisputable fact by a majority of the "educated public", doubtless to the delight of those like Richard Dawkins (*The Blind Watchmaker: Why the Evidence of Evolution Reveals a Universe without Design* [New York: W. W. Norton, 1986]), whose explicit and unabashed intention is to dethrone the Divinity of traditional theism. One hopes that books like Michael Denton, *Evolution: A Theory in Crisis* (Bethesda, Maryland: Adler and Adler, 1986), Phillip E. Johnson, *Darwin on Trial* (Downers Grove, Illinois: InterVarsity Press, 1993), and Michael Behe, *Darwin's Black Box: The Biochemical Challenge to Evolution* (New York: The Free Press, 1996) will help in exposing the common fallacy that transformism cannot be scientifically challenged. Their detailed examinations of the physical evidence against evolution, drawn among others from the disciplines of paleontology, comparative anatomy, and molecular biology, might be usefully studied as an empirical complement to the strictly metaphysical arguments presented here.

hension of the Real as Object by virtue of the Real as Subject; it is a matter of *theoria*, not theory, and rests upon the knowledge of like by like, when What truly is so becomes aware of Itself. The latter, theology, though not equally certain with respect to its mode of reception, is equally true in its content; for theological thinking— one must add, in its orthodox forms—is dependent on Revelation, and Revelation is intellection *quenched* as it were in the form of result: not knowledge, as Plato would say, but true opinion, orthodoxy. Opinion *per se*, however—whatever so-and-so happens to think at this moment about his more or less regular contacts with the more or less fluid domain of so-called physical "things"—is comparable to neither; and a thought dependent on the data of sense cannot but be an opinion.

The second reason is this. Like all scientific theories, transformist evolution (even supposing its validity) is limited to the strictly material or terrestrial plane, which is only a part of the cosmos, the least real and accordingly the least intelligible. As he attempts to explain what he sees, the Darwinian theorist neglects to remember that what can be physically sensed is not only much less than the whole, but less real than the "parts" of the whole which it is not, as the images of a dream are less real than the objects of waking perception. And yet this quite partial character of his hypothesis—and by "partial" I mean both "biased" and "incomplete"—is seldom if ever considered. Just the reverse: it is often assumed instead, not that all that is said about fossils applies in addition all the way up to the angels, though this would be more than absurd enough, but that there is no "up" in the first place, nothing besides matter at all. *De non apparentibus et non existentibus eadem est ratio*. The indefensible presumption of this materialism should be obvious.

A third observation naturally follows, and it concerns the reductionism implicit in the transformist cosmogony, its attempt to explain the more by the less. I have said that the scientific evolutionist often assumes that matter is all that exists, or matter-energy if a greater subtlety is required. But even when he does not so assume, or says he does not, the theory invariably leads its proponent to think that if there *is* something more or higher than physical substance, it can be approached only by way of the lower, and only as the product, result, or extension of processes and forces first apprehended, or inferred, at the empirical level. But if the lower explains the higher, then the lowest will explain most of all, and one is left to conclude, however preposterously, that something is derived from nothing, *quod absit.* It is as though the evolutionist had transferred the creationist cosmogony, uninformed by metaphysics, onto the material plane, together with the problems already considered, but without the theologian's God, and the result is doubly absurd: the creation of something *from* nothing *by* nothing. Logic itself compels one to see that "every productive cause is superior to that which it produces," and that "whatever is produced by secondary beings is in a greater measure produced from those prior and more determinative principles from which the secondary were themselves derived" (Proclus).

The fourth problem with the transformist position is that it mistakes temporal or chronological succession for ontological causation and so falls prey to the sophism *post hoc, ergo propter hoc.* Neither the metaphysician nor the theologian denies, or needs to deny, that the fossilized vestiges of various species of organic life appear in the geological record in an order of increasing complexity. Nor must he posit some extraordinary act of God by virtue of which that record was made to appear as it does by a miraculous "pre-fabrication", after the fashion of certain "scientific creationist" schemes. Instead he accepts the fact, though *a priori* and not by induction, that the various kinds of plants and animals have appeared successively over time, with the sensible manifestation of humanity coming near the end of the process. But appearance is one thing, and Reality quite another. To admit that reptiles roamed the earth before the appearance of mammals, or more precisely that they entered into the substance of physical bodies in advance of the mammals, does not entail the admission that the latter, by whatever temporal and biological channels one might wish to propose, therefore came *from* the former, though they may in a sense have come *through* them, as I shall later suggest. The metaphysician teaches instead, and quite the reverse, that the order in which the various species have been deployed over time, and thus the evolutionary sequence of their disclosure in matter as physical organisms, is just the opposite of their "original" order as archetypes or Divine ideas. Inasmuch as the world is the reflection of God, and inasmuch as reflections invert, this is precisely what one should expect—a kind of chiastic reduplication of the higher in the lower. It is essential of course to remember that the procession of the world out of God is eternal, hence in temporal terms both continuous and instantaneous. When it is said that man's appearance on earth toward the end of the cosmological "process" is the inverted reflection of his theological and metaphysical primacy in the Divine Logos, it should be understood that this primacy refers to the "spatial" super-ordination of human beings over animals in a purely ontological, static, and vertical sense, and not to their being an initial effect in a series of sequential creational acts. Man's primacy is that of the microcosm, of a "container" with respect to its "contents".

Fifthly, the Darwinian conception of evolution is utterly blind to the essential distinction of form from shape. By the word "species" it means to signify what many individual organisms have in common: a notional abstraction derived by generalization from particular facts and serving as a linguistic tool for classification. But these facts are all of an empirical, material kind and have to do solely with structural resemblances and other physical features susceptible to measurement and quantification. A species thus considered, *ab extra ad intra* and as it were by dissection, is therefore dependent on shape, even as shape is a function of surface. By "surface" we do not mean only the most outwardly external plane of an organism's solidity, the place where its skin (or some similar feature) meets the air, but also whatever relatively inward part of its physical substance could be exposed to the air, in fact or in principle, if our techniques were sufficiently refined. In this sense, what can be seen even through the most powerful of electron microscopes, or read on the screen of some other highly sophisticated detection apparatus, is still only "surface", and still therefore a matter of shape. Form is quite different. By "form" the metaphysician means to refer to that quality by virtue of which a physical object, whether living or not, transfers the attention of those who perceive it through itself and along a kind of ontological corridor up and into its celestial archetype. Form is liquid where shape is solid, though solids of course nonetheless display the various forms; form is transparent or diaphanous whereas shape is opaque; form insists that it not be confused with a surface. As shape is the place where empirical apprehension must necessarily stop on its way in the direction of being, form is the place where being willingly pauses on its way in the direction of knowledge. Form is on the "other side" of the existence of things from species and shape. Not at all derived from those things or dependent upon them, they are rather derived from it. It is what accounts for the qualitative existence of the animals or plants that possess it in common—whose commonality or similarity, however, far from being the measure or the standard of form, is as much an indication of the fragmentation and incompleteness of these particular creatures as it is a reminder of that which is whole, and which they seek to emulate, which is, once again, their form. A truly adequate cosmological explanation of the world is an explanation of forms and their hierarchical order, and of That which they wish to express. But all this, for obvious reasons, is quite beyond the expectations of a strictly empirical scientific method, the kind of method that has generated the transformist doctrine.

The sixth and final problem with this theory, the most fatal of all, is its failure to account for the mind of the theorist himself, and thus its suicidal self-contradiction. If the Darwinian is correct that the body of man, his brain included, like the bodies of the other animals, is the result of the operation of certain subordinate forces, whether physical, chemical, or biological; and if he is correct to assume that the mind is in some way a function of organic tissue and that thoughts are the result of electrical transmissions in the cerebrum: then precisely *because* of his claims he is *not* correct, nor could he be "correct" whatever he said. For from this point of view, no idea, including the theory of evolution, is true; none is conformed or adequate to the actual nature of things in a way that its competitors are not. If the transformist is right, all ideas, including his own, and thus all possible theories, are already equally conformed to the real, because they are all equally determined to be what they are by their respective biochemical histories. And so to repeat, *if* he is right he is *not*. Truth, like the mind which thinks it, requires a freedom from all conditioning, hence a freedom from all "horizontality", all physical process, and thus from the whole of nature, which is by definition a concatenation of effected causes and caused effects. Nothing is true—including the statement that "nothing is true"— unless the power of knowing within us "came down from Heaven", having proceeded out of the only completely unconditioned Reality: the Source of emanation and the God of creation. "In Him was life, and the life was the light of men."

In sum, the Darwinian theory of evolution reflects in various ways, and as considered in its several particulars, each of the more or less typical tendencies of modern critical thinking: it is by turns empiricistic, materialistic, reductionistic, historicistic, nominalistic, and relativistic. It is a veritable "hydra of heresies". And it is for this reason that the theory has been the object of so severe a reproach on the part of both the metaphysician and the orthodox theologian. Let us be clear. The evolutionism which must be repudiated is in no sense the claim that there has been a sequence in the appearance of the forms of life on this planet, nor that changes have occurred and continue to occur in the physical constitution of the various species of plants and animals, for these facts, to the degree that the data of sense can provide us with "facts", are amply attested by the geological record, by the techniques of radiometric dating, and by the observations of breeders.[6] What must be rejected instead is the attempt to explain such changes in strictly physical terms as if that explanation accounted not only for intraspecific differences between individual organisms, but for the existence and variety of species as such, and as if those species had developed *ex nihilo* from the simple to the complex, with the inanimate giving rise to the animate, the animate to the sentient, and the sentient to the self-conscious—the last of which must clearly be first in any intelligible causational series. Efforts to describe physical phenomena at the level of empirically observable causes alone are one thing, and are as useful in certain limited cases as they are unobjectionable. But it is quite another to insist, as the more hyperbolic of Darwinians often do, that an empirical explanation renders all other accounts of the same phenomena impossible, or that it can account in addition for the origin of non-empirical realities, or (worst of all) that it somehow proves that only such things exist as can be empirically measured. Herein lie the absurdities mentioned earlier.

*

The question remains, however, whether it is possible to formulate a view of the cosmos from an infrapersonal angle that is free

---

6 "Modern science is right when it describes the succession of geological periods, but not when it tries to describe the origins of life or of intelligence. Modern cosmology cannot be something other than geology, paleontology, and astronomy; and there exists not the least difficulty in combining them either with Semitic creationism or with Indo-Greek emanationism, for the simple reason that facts are always compatible with principles" (Frithjof Schuon, unpublished letter to the author).

from these several defects. Can the major forms of life in this world be described according to the temporal order in which their appearance occurs *per ascensum,* but so as to protect the metaphysical and theological truths of their "original Origin"—from and by God *per descensum?* Is it possible to envision the chronological order among the kingdoms on earth in such a way as to see, by a kind of noetic transposition, their ontological order in God?

In order for this question to be answered in the affirmative, it is essential that one first understand the purpose of cosmologies in general—perhaps we should say of *traditional* cosmologies or "natural sciences", lest they be confused with fields of study like astrophysics, which though it may purport to consider the cosmos as a whole suffers in obvious ways from the same limitations as the other empirical sciences. The traditional intention is basically this. Whatever shape it takes, and whatever the symbols or language it might employ, a true cosmology, by virtue of its conformity with metaphysical and theological principles, must be such as to make the *natural* a means of support for our awareness of the *supernatural,* whence nature proceeds and upon which it is permanently and perpetually dependent. A vision of the world which is not conducive to this intellection, and hence to the actualization of what is highest in man, simply cannot be admitted as valid. The strictly utilitarian tendencies of our age notwithstanding, the only good reason for an *infrapersonal* approach to the cosmos is that it might serve as an aid on the spiritual path toward the *transpersonal.* If it fails to do this, it has lost its very reason for being. Indeed this is the most fundamental danger with Darwinian evolutionism, that it causes a man to forget where he came from, thus abolishing his nobility.

Nevertheless, provided that this danger is kept in full view, the metaphysician and theologian need not *in principle* be opposed to the idea that the world is evolving. There is after all nothing etymologically wrong with the word "evolution", nor anything intrinsically absurd about every activity or process which it might be used to signify.[7] As we have seen, the transpersonal explanation is agreed

---

7 The term obviously had a pre-Darwinian history. The earliest English uses recorded in the *Oxford English Dictionary* are by two of the Cambridge Platonists, Henry More and Ralph Cudworth, whose understanding of the world and its creatures was quite far from transformist. And when Dr Samuel Johnson later wrote that "he whose task is to reap and thresh will not be contented without examining

that the cosmos becomes what it is through an "unwinding" or explication of What is already inside, which is "turned out" or *evolved* into what It is initially not, but can then be seen in. As long as the true nature of this original "inwardness", which is of course God in His immanent presence—"the dearest freshness deep down things"—is not debased by confusion with something created or relative; as long as we recognize that if matter can be said to evolve, it is only because of the presence within it of the Supreme Reality, which it expresses and seeks to return to; and as long as we realize that the world is not what the materialist thinks, but at once a symbol, a veil, and a channel: then an evolutionary cosmogony can be legitimately entertained, not simply as an acceptable theory, but as a genuine *theoria*, a vision opening up into metaphysical insight. If it is the case that the cosmos is a message sent by God from Himself to Himself, then we may expect the metaphysical truth of emanation, which considers the world as proceeding *from* God, and the theological truth of creation, which explains the world as fashioned *by* God, to be open to a corresponding scientific "truth", according to which the world is envisioned *on its way back to* God; and we would not be wrong in calling this last process an "evolution".

The distinctive features of an evolution consistent with God and conformed to His nature have already been anticipated indirectly, and by contrast, in our treatment above of the six Darwinian errors. The task now is largely one of organization and synthesis as we present a sketch, not of course of transformism, but of what might be called an "emanational evolution". The word *sketch* should be stressed. What follow are no more than leading thoughts, provocations and pictures, designed to suggest a possible line of reflection. There is certainly no intention of exhausting the topic, nor do we aim to anticipate all the conceivable objections. It should be emphasized in any case that the temporal terminology with which we are obliged to speak is not to be taken literally, for obviously God is not Himself subject to the developments here envisioned; only in

the *evolution* of the seed", it is clear he was speaking of how oaks come from acorns, and not men from the "lower primates". On the other hand, these philological data should not blind us to the fact that the connotation of the term has certainly changed since the eighteenth century, and that its user today must be extremely cautious lest he be thought a Darwinian.

seeming does the deployment of creatures "take time". The difficulty, of course, is that our discursive thought must by definition do
its work in that seeming and with the materials thus afforded.
Hence "stage and sequence are transferred, for clarity of exposition, to things whose being and definite form are eternal"
(Plotinus). This caveat must be kept in mind throughout.

By "emanational evolution" we mean two things: the deployment of form in matter as shape and the explication in the substance described by that shape of certain qualities or attributes. The
total process may be called emanational since both of its "stages",
both the deployment and the explication, proceed in the first place
per descensum from higher, immaterial planes. The cosmogony here
presented, though it offers a reason for the cosmos "from beneath",
thus begins "from above", in accordance with the stipulations
already established. More precisely, it begins by analogy and, for the
sake of *maieusis*, with the act of knowing, which itself proceeds also
from the higher to lower, "for everything that is known is known not
according to its own power, but rather according to the capacity of
the knower" (Boethius). And it begins by implication with that One
which is the principial Knower, and knowing Principle, wherever
the knowledge of being arises. The *metaphysical* scientist or cosmologist begins in this way with the Self as Subject.[8]

The Self is truly Subject only in its recognition of Itself as such,
and only in knowing Itself as the Knower It is, hence only to the
extent that It "becomes" an object as well. One says "as well", but of
course the Principle from which we are starting, and which these
altogether inadequate words are meant to evoke, is not an object *as
well as* or *in addition to* Subject. Rather than both It is neither,
though within this "non-dual" transpersonal primacy, It is nonetheless rather more like what is meant by a subject than by an object. If
we allow as a provisional means the language of change and transformation in the case of the strictly changeless and impassible, it
can be said that what was "once" purely Subject "becomes" Its own

---

8 No attempt is here made to defend the Platonism and Vedantism implied by
these assertions, though the substance of that defense is implicit in our reference
above to the sixth of the Darwinian mistakes. An explicit and extended consideration of the primacy of consciousness can be found in our book *The Form of
Transformed Vision: Coleridge and the Knowledge of God* (Macon, Georgia: Mercer
University Press, 1987).

object in the "midst" of Its act of knowing, for we are speaking of something whose very nature is such that It cannot but radiate outside of Itself, outside Its "initial" subjectivity, thus taking the form of "whatever is left"—that is, the object, in whatever modality one might wish to imagine. Of course, to speak in these temporal terms is only symbolic, as we have already noted. What are here being pictured as the stages of a process in time are in fact the multiple states of a single, non-temporal Essence.[9] The stages are no more than superimposed "translucencies" of varying colors through which one is intended to glimpse, as "through a glass darkly", the infinity of that Essence.

It may prove helpful to make use of an image. Consider a point. The geometrical point is after all the closest of all mathematical forms to the nature of the Subject and to the root of intellection. Altogether independent from objects, even from the dot that serves as its representation, but which is already more extended than an actual *punctum*, the point is pure inwardness, uncompromised by any equivalent externality to which it might otherwise be thought to correspond. It is "an inside without because prior to an outside".[10] And yet again like the Subject, the point tends to search for itself, to spill over the edges of its invisible and dimensionless essence into the dimensions of the things that are seen. In seeking itself, it produces in the first place a line, a figure we may mistake for a collection of discontinuous "points", but which (as Zeno implied) is more like an effort or energy suspended between them, or better *expended*

9 "The concept of multiple states permits us to envisage all these states as existing simultaneously in one and the same being, and not as only able to be traversed successively in the course of a 'descent' which supposedly passed not only from one being to another, but even from one species to another" (René Guénon, *The Multiple States of Being*, trans. Joscelyn Godwin [Burdett, New York: Larson Publications, 1984], 73).

10 This locution is borrowed from Coleridge, as is the point as image of the Subject. Suggestions for several additional features of our "sketch" can be found in Coleridge, *Biographia Literaria*, ed. James Engell and W. Jackson Bate (Princeton: Princeton University Press, 1983), Ch. 12; and in "Hints Toward the Formation of a More Comprehensive Theory of Life", *Selected Poetry and Prose of Coleridge*, ed. Donald A. Stauffer (New York: Random House, 1951). Similar employments of the geometrical point can be found in St Clement of Alexandria, and René Guénon makes use of the image throughout his *Symbolism of the Cross*, trans. Angus MacNab (London: Luzac, 1975).

by one such point—stretched as it were between itself in the form of subject and itself as found in the object. This line, one might say, is the line of knowledge, connecting the Self or Principle with each of Its innumerable objectifications.

But the point is not "satisfied" with the line, which actualizes its possibilities in but a single direction. It searches for breadth in addition to length, projecting itself at right angles to its original motion. In this way, the plane is "evolved", which in metaphysical terms is the demarcation of a particular level of being—in this case, below the Divine but above the visibly human—a plane which provides for the "clothing" of the as-yet-immaterial entities that dwell upon it. No longer quite pure in its initial inwardness, the point as expressed in this dimension is nevertheless still intangible, invisible, and unresisting—still fluid because having no depth or physical substance, containing as yet nothing external in any material sense. Here is the field, in traditional cosmological language, of the subtle body and its world, the *mundus imaginalis*.

And yet, still unquenched in its thirst, the Subject as point become line and then plane must "explode" once again, in this case into the solid.[11] It is now for the first time that the energy of ontological knowing becomes coagulated or fixed, as what was before a "gas" and then a "liquid" now becomes "solid", possessed of three dimensions. What had been an idea in the "celestial" world, an ontological ray *projected* by the Supreme Knower, and what then was *condensed* on the "intermediate" plane, is now *frozen* by the dimension of depth as it enters the "terrestrial" order, which is the familiar

---

11 The word "explode" is deliberately used, and it could in fact be repeated below with respect to each of the "stages" and forms of existence envisioned, in order to denote the discontinuity between various levels and the suddenness with which they arise. Martin Lings has pointed out that the teaching of Jalal ad-Din Rumi concerning rebirth "from vegetable to animal and from animal to man", analogous in certain respects to the sequence we are about to discuss, has sometimes been misinterpreted as an anticipation of Darwinian "evolutionism", in spite of the fact that in Rumi "there is no gradual development but a series of sudden transformations", and even though "the mineral, vegetable, animal, and human states are envisaged"—both in Rumi and here—"as already existing and fully developed. The evolution in question is that of a single being, from the lowest to the highest states, from the periphery to the centre" (Lings, *The Eleventh Hour: The Spiritual Crisis of the Modern World in the Light of Tradition and Prophecy* [Cambridge, England: Quinta Essentia, 1987], 28n).

world of material objects and empirical perceptions, the world studied by the physical scientist and Darwinian.[12] It is of course here, at the level of matter, that the evolution of species is usually assumed to begin, by which "evolution" is normally meant: first the production of life from inorganic substances, and then the development of the various living species from simple to complex, culminating in man. From the point of view of pure metaphysics, however, the true development must be understood very differently: not to the exclusion of all that exceeds the material order, but in full view of the "higher", and as an extension or prolongation of the earth's pre-material history.[13]

This "history", the emanational sequence by virtue of which the physical plane proceeds by way of the subtle or animic order from out of its principial Source, is sometimes referred to as an "involution", a word which calls attention less to the metaphysical movement *in divinis* from within to outside—to evolution *per se*—and more to the cosmic receptacle into which the movement enters, into which the Subject is incarnated or emptied, "Who being in the form of God made Himself of no reputation, and took the form of a servant, and was made in the likeness of man" (Phil. 2:6-7). From another point of view, however, the one which is here being stressed, the entire process, whether in its involutionary or evolutionary "stages", and whether outside or inside the material plane—deployment or explication—is all of one piece: it is an evolution or "opening out" of none other than God throughout its entire extent. For it is one and the same Subject—or, to make use of our image again, the very same point—which accounts for the whole, for all of the degrees of Reality and all of their innumerable contents. Of

---

12 The terms "celestial", "intermediate", and "terrestrial" have been used by Huston Smith to designate the degrees of Reality below the "Infinite", below (that is) what we are calling the Subject and depicting by means of the point. See his *Forgotten Truth*, Chapter 3.

13 "A minus always presupposes an initial plus, so that a seeming evolution is no more than the quite provisional unfolding of a pre-existing result; the human embryo becomes a man because that is what it already is; no 'evolution' will produce a man from an animal embryo. In the same way the whole cosmos can only spring from an embryonic state which contains the virtuality of all its possible deployment and simply makes manifest on the plane of contingencies an infinitely higher and transcendent prototype" (Frithjof Schuon, *Understanding Islam* [World Wisdom, Bloomington, Indiana, 1998], p. 130n).

course to see that this is so requires a method of investigation and a perceptual sensitivity quite different from the ordinary empirical orientation. Things must be approached from *within*, with a view to their forms, not to their shapes. We have said that one of the problems with transformist evolution is its inadequacy to the full range of the cosmos, that it looks at only a *part* of the whole. But the problem is also that it looks only *at* a part of the whole. Its approach to the world is therefore bound to eclipse the metaphysical transparency of natural forms, which are intended by God to be His self-expression, the expression of "even his eternal power and Godhead" (Rom. 1:20), but which can be seen as such only if men look not at but *along* or perhaps even *through* the creatures around them.

If they were so to look, what they would see is that the production of solids from their source in the point is not the end of the story. The radiantly explosive energy of the Subject is not yet exhausted, nor can it be. But having as it were no more room to expand, all the dimensions being filled with Its presence, It cannot but turn in Its movement toward the center of the things it has made. This center must certainly not be confused with some sort of spatial position, half-way between the front and the back, or the top and the bottom, of a particular solid entity. It is rather the "inside without because prior to an outside" of the point itself, into which, as into Itself, the Subject now proceeds. The resulting introsusception marks the beginning of an act of return, of remanation and recapitulation. And yet it is also "more of the same"— simply another manifestation of the Self's unfathomable plenitude.

Having then, to repeat, no more room for itself, in "need" of expansion but lacking an adequate space, the point undertakes something "new" and begins to move along a novel course. It begins to display in the solids once made a range of qualities or attributes, unveiling itself in the *how* not the *what*, in the kind not the fact, of the already existing material objects. Thus no sooner are the solids crystallized than, by virtue of the same pre-material emanational energy that first produced them, there begins to burgeon inside a series of "higher dimensions". An interiorizing action is initiated, according to which the manifestations of the Subject become increasingly central to the creatures in question.

At first the point, having just completed its third and final "spatial" deployment, presents itself in a uniform way throughout a particular object, equally in length, in breadth, and in depth, and in the form of a special property, as for example in the attractive power of the lodestone or the crystalline structure of the diamond. Such a property is everywhere all that it is, as much in the parts as in the whole, a piece of a given substance having "as much" of the quality—to use the uselessness of such language—as any other. The Source of the substance, for those who have eyes to see, is thus written upon its very surface.

But the surface of things is far too confining for the Subject. It needs the amplitude afforded by Its own infinity, and soon becomes "discontent", diving below the surface of physical objects, going "indoors", and there presenting itself in the guise of a process. Here is the beginning of life in its organic expression and as unfolded on the inside of matter. What one immediately notices in looking along such a process, however—along a vital or physiological function—is that a certain *specialization* among the parts of the whole has been introduced. Unlike a mineral, not all of a living being is all that it is all the way through. Certain aspects, even when they are not as clearly distinct as the individual organs of the higher species, assume a certain priority as providing the channels or openings through which life may enter the entire being. These aspects are to the substance of a living creature what the plane is to the solid; they are its *conditio sine qua non*—in it, but not of it. Quite the contrary, it is dependent on them. But more to the purpose, where mineral properties had been too restrictive, the more inward character of biological process provides the Subject with additional opportunities for expansion, as is evident in growth. Although the manifestation of the Subject in organic life has become in a sense itself restricted to certain more or less specialized aspects of an entity, and in a way that the properties of a diamond are not, these aspects are actually open along a metaphysical passageway to far wider possibilities than inorganic matter *per se* can allow for, and it is thus that there enter into the being the "uncreated energies" that are necessary for the unfolding of its life. It is in this sense that the plant kingdom, though emerging later in the temporal sequence, may be said to be higher than the mineral, for the plant provides a greater degree of openness to its source.[14]

Now we encounter yet another development, a further "evolution" of the point. Although the plant is able to grow and blossom in a way that the mineral cannot, it remains subject to limitations, which the Subject must still overcome, limitations visibly expressed in its attachment to the earth. If unrestricted emanation is to continue, therefore, an altogether different access must be provided through which the point might be enabled to pass into an even more resplendent qualitative expression. Here, it would seem, is the metaphysical explanation for the birth of sentience and the power of locomotion, the two most telling features of the kingdom next to develop, that of the animals. An even more pronounced *interiorization* has here taken place, for while an animal's movement is already a sign of deeper capacities evident even to a purely empirical outlook, the power of conscious response which gives rise to this movement, and which can be seen (once again) only by "looking along" the locomotion, is even further *within* than the power of life, and yet, precisely because it is further inside, it is correspondingly closer to an exteriorization or evolution on the other side of the creature, the side opposed to the facet discerned by the purely sensory eye—even as form is on the further, but shape on the hither, side of material objects. As the point strives to make room for itself in a space that is already exhausted, channels are established through which there can flow the greater intensity of sentience. It is in this way that the flower is opened to the butterfly, a blossom released from its stem.[15] Nevertheless, this release or liberation could not have been accomplished by the plant on its own.

---

14 A qualification to this principle is in order, however, since the "metaphysical transparency" (Frithjof Schuon) of a given creature depends not only upon the plane of being it occupies, but upon its relative "centrality" on that plane—its proximity to the point at which that plane is intersected by the *axis mundi*—not to mention other imponderables. While plants may in general be "higher" than minerals, because of the organic life that is in them, a precious gem remains a more lucid theophany than a weed. The same qualification must be applied to the relative positions in the great chain of being of plants and animals, and of animals and man. A noble animal, like an eagle or a lion, is "more Divine" than a human being who lives below himself. We do not in any case mean to propose a rigid system.

15 "The insect world, taken at large, appears as an intenser life, that has struggled itself loose and become emancipated from vegetation . . . . We might imagine the

The flower has served as a portal for creative, causative Power, but is not itself that cause. Even as apart from the plane the line would remain invisible, in proportion as the plane apart from the solid would be intangible, so in the absence of life would consciousness be hidden from the organs of sense. On the other hand, the plane could not exist at all apart from the line, nor the solid without the plane, nor any dimension were it not for the point; and were it not for the Subject, neither matter, nor life, nor sentience would ever have been deployed or explicated—Subject through matter as life, and Subject through life as sentience.

It is worth pausing to emphasize the radical difference between this view of the sequence of forms and that of the Darwinian or transformist theory. The metaphysician is *not* saying that matter evolves into life, or life into sentience. Nor, on a smaller scale, do amphibians evolve into reptiles, or reptiles into birds; Galapagos finches are always finches, and peppered moths are always moths.[16] The only evolution is that of the point, which is the Divine Self as Subject. The forms of existence through which It "passes", in a strictly non-temporal and instantaneous way, do not themselves change, for they are the unalterable images of celestial ideas—the distinct and immutable shadows cast by the Divine Sun as It shines upon the eternal archetypes of Its myriad creatures.[17]

---

life of insects an apotheosis of the petals, stamina, and nectaries round which they flutter, or of the stems and pedicles, to which they adhere" (Coleridge, "Theory of Life", 594).

16 The variations within certain populations of finches, first noted by Darwin in the Galapagos Islands during his 1831-36 voyages on the H.M.S. *Beagle*, are well known. As for this particular species of moth, "Kettlewell's observation of industrial melanism in the peppered moth (Biston betularia) has been cited in countless textbooks and popular treatises as proof that natural selection has the kind of generative power needed to produce new kinds of complex organs and organisms", even though in this case there was never any change of the kind alleged by transformists (Johnson, *Darwin on Trial*, 176).

17 "Evolution is the unfolding of a given virtuality and not the passage of a given possibility to a quite different possibility" (Frithjof Schuon, unpublished letter to the author). "The form of a peripheral being, whether it be animal, vegetable, or mineral, reveals all that the being knows, and is as it were itself identified with this knowledge; it can be said, therefore, that the form of such a being gives a true indication of its contemplative state or dream. . . . Needless to say, the object of knowledge or of intelligence is always and by definition the Divine Principle and cannot

There is obviously a final chapter, a final stage to be sketched, and this is the case of man himself, and hence of the mind which theorizes with a view to the truth: man, in whose various levels of being there is uniquely exhibited, emerging *through* the sentience he shares with the animals, the life he shares with the plants, and the body he shares with the minerals, an incomparably different quality, his consciousness—one should perhaps say "potential" or "virtual" consciousness—of Self.[18] Here the centripetal, intropulsive, and interiorizing tendency of the point has "at last" *caught up* to itself, having step by step provided ever more extended possibilities for irradiation, "until" the level is reached where it "becomes" the field of its very own motion and expression. The circle has been closed; the goal is attained. Intellection has fulfilled its reason for being and has realized in the only fully adequate way the plenitude of its unconditioned freedom, within the limitless "space" of its very Self. For to say man, the last of the species to appear on the earth, is to say capacity for the Absolute and the Infinite, for the Supreme Reality, since it is only in man, of all creatures, that the Subject is able to find a sufficient accommodation for Its full intensity. No mere bodily, three dimensional entity as such can hope to contain the *fons et origo* of all things, nor do the "dimensional" qualities, whether property (as a recapitulation of the solid), life (as a recapitulation of the plane), or sentience (as a recapitulation of the line) ever more than approach It. Its energy is such that It can be fully manifest only in a mode which is equally infinite and therefore equally without dimension—equally like a point. But this is pre-

be anything else, since It is metaphysically the only Reality; but this object or content can vary in form in conformity with the indefinite diversity of the modes and degrees of Intelligence reflected in creatures" (Frithjof Schuon, *The Transcendent Unity of Religions* [Wheaton, Illinois: The Theosophical Publishing House, 1984], 56).

18 Notice the emphasis on the preposition *through*, not "from". Man's consciousness of himself, his power of self-reflection, can obviously come only *from* the Self, on pain of the contradiction referred to earlier in considering the sixth of the Darwinian errors. Failure to attend to this crucial "prepositional" difference can result in a kind of "optical illusion", which is perhaps at the root of that impossible compromise popularly known as "theistic evolutionism", whose proponents contend that God somehow manages to create the world by means of the process described by Darwin. It should be clear by now that the *metaphysical* or *emanational* "evolution" here envisioned has nothing in common with this absurdity.

cisely what Self-consciousness is, what intellection and *noesis* are: an unbounded instantaneous inwardness, which, like the attributes of life and sentience, is not at all *of* the human body, but which unlike these lower qualities is not even *in* it, because not expressed even as it were on its "edge", as the plane and the line can be seen on the sides and the corners of solids. "Word as He was, so far from being contained by anything, He rather contained all things Himself" (St Athanasius).[19] The act of ontological knowing issues from a central and hidden point so concentrated as to exceed altogether the reach of the body and, hence, the all too limited means of empirical research. Although as this sketch has been intended to show man's intelligence can be said to evolve *through* what is beneath it in the order of being, it evolves even so, and can only evolve, from the higher and indeed as the Highest. What is length, breadth, and then depth on the way "down" to the earth becomes triply and inversely displayed on its way "in", and in a certain sense "up", as the qualities of physical substance—in property, process, and sentience—the prematerial construction of matter being mirrored in the succession of creatures. And yet none of the dimensions would be what it is, nor would the dimensional qualities be recognized for what they are, if it were not for the point, the sovereign Subject, which is both beginning and end.[20]

*

As announced at the outset, the aim of this article has been to provide an account of the world *per ascensum* which is in full con-

---

19 The saint adds that "man, enclosed on every side by the works of creation and everywhere beholding the *unfolded Godhead of the Word*, is no longer deceived concerning God" (our italics).

20 "When all is said and done, there are only three miracles: existence, life, intelligence; with intelligence, the curve springing from God closes on itself, like a ring which in reality has never been parted from the Infinite" (Frithjof Schuon, *Light on the Ancient Worlds*, 42). "As regards manifestation, it may be said that the 'Self' develops Its manifold possibilities, indefinite in their multitude, through a multiplicity of modalities of realization, amounting, for the integral being"—represented here by the point—"to so many different states, of which states one alone, limited by the special conditions of existence which define it, constitutes the portion or rather the particular determination of that being which is called human individuality" (René Guénon, *Man and His Becoming According to the Vedanta*, trans. Richard C. Nicholson [New York: The Noonday Press, 1958], 29-30).

formity with the view *per descensum* propounded by metaphysics and theology; and we have therefore been obliged to insist throughout that none "hath ascended up to Heaven, but He that came down from Heaven, even the Son of Man, which is in Heaven" (John 3:13)—hence that no infrapersonal or evolutionary explanation of the world can have the slightest value unless it acknowledges, and repeatedly stresses, the absolute supremacy of God, our ultimate Source.[21] Nonetheless we have also assumed that the sequence in the terrestrial appearance of certain forms of existence is not without a symbolic message, and that reading that message in the light provided by traditional authorities might help to point certain of our contemporaries, who wish to make so much of the "facts" and who often falsely suppose them to be incompatible with a Divine explanation, *back* and *up* to principles. For truths are always consistent with Truth.

21 Though far less has been said of a specifically "personal" nature or in dogmatic terms, and far more along metaphysical lines, the theological bases for an acceptable interpretation of evolution should be deducible from this and other traditional Christian teachings alluded to at various points throughout this article. Further support can be found in the dogma of the early Church regarding the Blessed Virgin as the *Theotokos* or "Mother of God", and in the metaphysical insight it affords concerning the substance of *materia prima*, a substance which is not unconnected (when rightly considered) to that "material matter" through which the "message" sent by God from Himself passes on its way of return. Even as that Man was born of Mary in Whom was "recapitulated the ancient making of Adam" (St Irenaeus)—that is, the eternal emanation of primordial man *in divinis*—the Virgin being the earthly expression of the "fecundation latent in eternity" (Eckhart): so there proceed out of matter, and by virtue of the "dimensional" elaborations described above, the forms of the minerals, plants, and animals, which were likewise "already" eternally made, and *through* the material "coagulations" of which the idea of Man is enabled to move in its "progress" towards full disclosure on earth. As St Anselm writes, "The Holy Ghost and 'the power of the Most High' wonderfully begat a Man from a Virgin Mother. Thus with respect to the others it lay in Adam, that is, in his power, that they should have being from him, but with respect to this Man it did not lie in Adam that He should exist in any way, any more than it lay in the slime that the first man, who was made from it, should come from it in a wonderful way, or in the man, that Eve should be of him, as in fact she was made. But it did not lie in any of them, in whom He was from Adam to Mary, that He should exist. Nonetheless, He was in them, because that from which He was to be taken was in them, just as that from which the first man was made was in the slime, and that from which Eve was made in him. He was in them, however, not by the creature's will or strength, but by the Divine Power alone."

Whatever the benefits of this reading, and quite apart from the question of whether it might prove as convincing to the theologian as to the metaphysician, it is worth emphasizing once again that the picture presented here is precisely a "picture"—no more than a sketch, a possible *darshana*. Simply because a certain vocabulary and particular images have been used as maieutic means, no one should imagine that all others have been thereby excluded. We readily admit that this cosmogony does not, cannot, and need not stand opposed to other equally effective "impersonal" visions, provided of course that they are equally subject to the irrefragable truths of pure metaphysics, concerning which there can be no compromise. It is to be understood in any case that the sufficient reason for having a view of the cosmos is *not* that it might correspond in some more or less mathematical way to the world as it is in itself, which is as far beyond perfect comprehension as it is below perfect being. The aim instead is to provide various *keys* or *supports* for intellection, that uncreated power of knowing What is and being What knows by which we are enabled to transcend this world altogether—though then only in seeing, through a prayerful looking *along* it, that this same world is "already" the kingdom of God.

Adveniat regnum tuum, fiat voluntas tua, sicut in caelo et in terra.

8

The Nature and Extent of Criticism
of Evolutionary Theory

Osman Bakar

In this essay, we will look into the existing body of criticisms which have been brought against the modern theory of evolution; we will investigate the nature and extent of these criticisms and conclude with an evaluation of their meanings and significance and the possible impact they will have on the future development of the theory.

Before we proceed to identify the above body of criticisms, we need to clarify the meaning of the precise idea or concept that is being criticized since the term *evolution* has been used to convey different meanings and connotations. Herbert Spencer, for example, who is considered the first great evolutionist and who gave the word evolution its modern connotation in English, used the word in two different senses in his essay *The Development Hypothesis*[1] which appeared in the *Leader* between 1851 and 1854, that is several years before the publication of Darwin's *The Origin of Species*. In this essay as well as in his later work *The Principles of Biology*, Spencer describes both the development of an individual adult organism from a mere egg and phylogenetic transformation of species as processes of evolution.[2] This usage of a single term, namely evolution, to describe two altogether fundamentally different processes has generally been avoided by today's scientists. But the possibility of confusion

---

1 This essay was reprinted in *Essays: Scientific, Political and Speculative* (London, 1868). In it Spencer asks why people find it so very difficult to suppose "that by any series of changes a protozoon should ever become a mammal" while an equally wonderful process of evolution, the development of an adult organism from a mere egg, stares them in the face. See Peter Medawar, *Pluto's Republic*, Oxford University Press (1982), p. 211.

2 See H. Spencer, *The Principles of Biology*, revised ed., London (1898), first volume.

remains because the term, though now restricted to one process alone, is still used differently by different sections of the scientific community. As pointed out by Sir Peter Medawar, the distinguished British biologist who was awarded the Nobel Prize for Medicine in 1960, biologists who use English as a scientific language never use the word "evolution" to describe the processes of growth and development because to do so would be confusing and misleading.[3] Among French scientists generally, however, it is the word *evolution* which is used to describe biological transformations within a particular species in adapting itself to a changed set of natural conditions while the supposed change of one species into another through natural agencies and processes is denoted by the term *transformism*.[4] It is in the sense of this transformism that we are using the term evolution here. And we are adopting this term instead of the word transformism precisely because, as pointed out by Professor S. H. Nasr, it contains a more general philosophical meaning outside the domain of biology not to be found in the more restricted term transformism.[5] Indeed, it will throw much light on the historical origin of the idea it conveys and its conceptual relationship with certain philosophical ideas that were dominant at the time of its formulation and this is of great relevance to our present discussion. In this essay, it is with the criticisms of the idea of evolution in the sense of transformism and its various implications that we concern ourselves.

More than a century after Darwin's publication of *The Origin of Species*,[6] opposition to the theory of evolution still continues and in

3 Medawar, Peter, *op. cit.*, pp. 215-216.

4 On the insistence of some scientists on a careful distinction between evolution and transformism, see M. Vernet, *Vernet contre Teilhard de Chardin*, Paris (1965).

5 See S. H. Nasr, *Knowledge and the Sacred*, Crossroad, New York (1981), p. 249.

6 *The Origin of Species* appeared on 24th November 1859 in an edition of 1,250 copies, all of which were sold on the first day. See Paul Edwards, ed., *The Encyclopedia of Philosophy*, Macmillan & Free Press, New York (1967), vol. 2, p. 249.

This extraordinary enthusiasm shown toward *The Origin* can only mean, and this is generally recognized now, that the idea of organic evolution was already widely discussed before *The Origin*. For a detailed inquiry into this pre-*Origin* discussion of organic evolution, see for example Arthur O. Lovejoy, "The Argument for Organic Evolution before *The Origin of Species*, 1830-1858," in B. Glass, O. Temkin, and W. L. Straus, eds., *Forerunners of Darwin*, 1745-1859, Johns Hopkins Press, Baltimore, 1968 edn., Chapter 13, pp. 356-414.

fact has been more widespread in the past several years. What is the nature of this opposition? There are many evolutionists who would like us to believe that whatever opposition there has been has come solely from the non-scientific quarters especially those who have their religious views and interests at stake. That such belief actually prevailed in the minds of most people for quite a long period of time, and is still widely held, is due mainly to the evolutionists' vast and well-established propaganda machine which ensures that no potential scientific opposition be given the opportunity to gain a foothold in the scientific establishment.

Now that the dissent and opposition within the scientific rank is too widespread to be ignored or contained, certain evolutionists are quick to justify the present state of controversy surrounding evolutionary theory as a natural consequence of the most extraordinary attention that biologists have given to the theory in nearly fifty years and also as reflecting a more critical acceptance of the theory on their part in contrast to the complacency of their predecessors.[7] Whatever justifications evolutionists may wish to advance, the fact is that today there are many scientists who oppose the theory of evolution on purely scientific grounds and in turn argue for the need of a positive alternative, namely a non-mechanistic explanation of the origin of life.[8]

More than fifteen years ago, the fact that there was a widespread dissatisfaction with evolutionary theory was already admitted. Sir Peter Medawar whom we have mentioned earlier, in his opening remarks as chairman of a symposium entitled *"Mathematical*

7 One such recent work which attempts to explain the meaning and significance of the present state of controversy in evolutionary biology is Niles Eldredge, *The Monkey Business: A Scientist Looks at Creationism*, Washington Square Press, New York, 1982. For example, he says, "Today, though chaos is too strong a word, there is definitely dissent in the ranks. Few biologists agree as completely and complacently as they did that short time ago. . . . The unusual thing about evolutionary biology is not its current state of flux. If anything was unusual, it was perhaps the period of quiescence and agreement from which evolutionary biology is only now beginning to emerge." p. 52.

8 One of the most recent additions to the list of scientific pleas for a non-physical, non-mechanistic explanation of the origin of living organisms is a work by Richard L. Thompson entitled, *Mechanistic and Nonmechanistic Science: An Investigation Into the Nature of Consciousness and Form*, Bala Books, New York (1981). Thompson is a mathematician and research scientist in mathematical biology.

Challenges to the Neo-Darwinian Interpretation of Evolution" held April 25 and 26, 1966 at the Wistar Institute of Anatomy and Biology, Philadelphia, said: "There is a pretty widespread sense of dissatisfaction about what has come to be thought of as the accepted evolutionary theory in the English-speaking world, the so-called Neo-Darwinian theory."[9] He identified three main quarters from which this dissatisfaction came: scientific, philosophical and religious.[10] To these we would add another important category of criticisms, namely the metaphysical and cosmological, which must be distinguished from the philosophical[11] and without which no study on contemporary opposition to evolutionary theory is complete. We consider these latter criticisms to be of greatest importance because they were missing in the original debate on evolution due to the eclipse of the metaphysical tradition in the Western intellectual firmament in the nineteenth century. In the absence of authentic metaphysical knowledge particularly pertaining to nature, and with nineteenth-century European theology unable to provide satisfactory answers to the problem of causality, the theory of evolution appeared to Western man then as the most plausible and rational explanation of the origin and diversity of life.[12] We now

9 P. S. Moorhead and M. M. Kaplan, eds., *Mathematical Challenges to the Neo-Darwinian Interpretation of Evolution*, p. XI. Quoted by A. E. Wilder-Smith, *The Creation of Life*, Wheaton, Illinois (1970), p.37.

10 Wilder-Smith, A. E., *op. cit.*, pp 37-38.

11 "Metaphysics is a science as strict and exact as mathematics and with the same clarity and certitude, but one which can only be attained through intellectual intuition and not simply through ratiocination. It thus differs from philosophy as it is usually understood. Rather, it is a *theoria* of reality whose realization means sanctity and spiritual perfection, and therefore can only be achieved within the cadre of a revealed tradition." S. H. Nasr, *Man and Nature*, Unwin Paperbacks, London (1976), p. 81.

12 "The understanding of metaphysics could at least make clear the often forgotten fact that the plausibility of the theory of evolution is based on several non-scientific factors belonging to the general philosophical climate of eighteenth-century and nineteenth-century Europe such as belief in progress, Deism which cut off the hands of the Creator from His creation and the reduction of reality to the two levels of mind and matter. Only with such beliefs could the theory of evolution appear as 'rational' and the most easy to accept for a world which had completely lost sight of the multiple levels of being and had reduced nature to a purely corporeal world totally cut off from any other order of existence." S. H. Nasr, op. cit., p. 125.

take a closer look at each of these types of criticisms and investigate to what extent the ideas embodied in them are being discussed within the academic community.

We begin with a survey of the historical origin and development of metaphysical criticisms of evolution. In his Gifford lectures presented in 1981, the first ever by a Muslim scholar, Professor Nasr conveys one important fact about the nineteenth century: it marks the peak of the eclipse of metaphysical tradition in the West. What rays of metaphysical light there were, associated with such names as Thomas Taylor, Goethe, Blake and Emerson, for one reason or other never succeeded in penetrating through the highly secularized philosophical and scientific layer enveloping the mind of Western man.[13] In reality, therefore, what characterized the nineteenth-century debate on evolution was the absence of its metaphysical dimension. But many exponents and defenders of evolution think otherwise. In their view, one of the achievements of Darwinian evolution was to break the hold on biological thinking of such metaphysical ideas as the immutability of species, divine archetype, creation and design or purpose in Nature, ideas which permeated pre-Darwinian biology.[14] It is true that all these ideas are contained in the teachings of traditional metaphysics. But these ideas also belong to popular theology. Between the metaphysical and the theological understandings of these ideas, there are significant differences whether it is in Islam or in Christianity. When these ideas were attacked by various quarters in the nineteenth-century West, their true metaphysical meanings were no longer in currency. The attack was therefore mainly directed toward the popular theological formulations of those ideas.

Take, for example, the idea of creation. What evolutionists have severely attacked is the theological conception of *creatio ex nihilo* (creation out of nothing). Metaphysicians understand the idea of creation differently. They refer to it as creative emanation. (A brief discussion of this important metaphysical idea is given below). Here there is no question of having to make a choice between creation *ex nihilo* and creative emanation. Both are true but at different levels.

---

13 See Nasr, S. H., *Knowledge and the Sacred*, pp. 97-99.
14 Paul Edwards, ed., *The Encyclopedia of Philosophy*, p. 303.

As pointed out by Frithjof Schuon (see below), creative emanation is not opposed to creation *ex nihilo.* In fact, the metaphysical conception of creative emanation explains the real meaning of *ex nihilo.* Both ideas are meant to fulfill the different needs of causality among different types of "mentality" found within a religious community. Within the religious world-view, the idea of creative emanation proved to be more attractive or satisfying to the scientifically and philosophically minded than the idea of creation *ex nihilo* in its theological sense. This is certainly true in the case of Islamic civilization. In that civilization many philosopher-scientists, apart from the Sufis, adopted emanation as the philosophical basis for the explanation of the origin of the universe and the emergence of different qualitative forms of life.

What about the idea of evolution itself? This question is answered by Martin Lings:

> The gradual ascent of no return that is envisaged by evolutionism is an idea that has been surreptitiously borrowed from religion and naïvely transferred from the supra-temporal to the temporal. The evolutionist has no right whatsoever to such an idea, and in entertaining it he is turning his back on his own scientific principles.[15]

Very few people today realize that the idea of evolution originally belonged to metaphysics. But in the nineteenth-century West, as we have previously stated, metaphysical ideas, including the idea of evolution, had all been emptied of their true metaphysical content through a long process of secularization. The evolutionary chain of living organisms in post-Darwinian biology is none other than the secularized and temporalized version of the traditional metaphysical doctrine of gradation or the "great chain of being" of the Western tradition. The whole set of "metaphysical" ideas, which are collectively referred to as creationism by some historians of science,[16] were understood then and have been understood ever since solely at the popular, theological level. Thus the true nature of

---

15 Martin Lings, "Signs of the Times" in *The Sword of Gnosis*, ed., Needleman, J. Baltimore (1974), p. 114.

16 See Gillespie, Neal C., *Charles Darwin and the Problem of Creation*, University of Chicago Press, Chicago (1979), Chapter 1.

the debate between evolution and creationism in the nineteenth century was anything but metaphysical.[17]

Metaphysical Criticisms of Evolution

What can properly be called metaphysical criticisms of evolution first appeared in the early part of this century in the writings of a small group of metaphysicians in the course of their presentation of the traditional doctrines of the Orient.[18] The first as well as the central figure most responsible for the presentation of these doctrines in their fullness was René Guénon (1886-1951), a Frenchman and a mathematician by training. His first book was published in 1921 and entitled *Introduction générale à l'étude des doctrines hindoues (General Introduction to the Study of the Hindu Doctrines)*. This was the first full exposition of the main aspects of traditional doctrines. A complete guide to René Guénon's intellectual career and works during the next thirty years was provided by another eminent metaphysician, Ananda K. Coomaraswamy (1877-1947) in an essay entitled *Eastern Wisdom and Western Knowledge*.[19]

Coomaraswamy, born of a Singalese father and an English mother, was a distinguished geologist before his conversion to traditional metaphysics. At twenty-two he contributed a paper on "Ceylon Rocks and Graphite" to the *Quarterly Journal of the Geological Society* and at twenty-five he was appointed director of the Mineralogical Survey of Ceylon. A few years later he was awarded the degree of Doctor of Science by the University of London for his work on the geology of Ceylon.[20] Like René Guénon, he also produced numerous articles and books on metaphysics and cosmology

---

17 *Editor's note:* "In reality, the evolutionist hypothesis is unnecessary because the creationist concept is so as well; for the creature appears on earth, not by falling from heaven, but by progressively passing—starting from the archetype—from the subtle to the material world, materialization being brought about within a kind of visible aura quite comparable to the 'spheres of light' which, according to many accounts, introduce and terminate celestial apparitions." Frithjof Schuon, *From the Divine to the Human* (World Wisdom, Bloomington, 1982), p. 88.

18 Nasr, S. H., *op. cit.*, p. 100.

19 Coomaraswamy, Ananda K., *The Bugbear of Literacy*, Perennial Books, Bedfront, Middlesex, Chapter IV, pp. 68-79, (1979 edn.)

20 *Ibid*, p. 8.

which in many respects complemented the works of the former.[21] Through his writings, Coomaraswamy played a great role in reviving the traditional point of view. Professor Nasr, in his study of the history of the dissemination of traditional teachings in the West during this century, considers the task of the completion of the revival of traditional metaphysics to have been accomplished through the writings of Frithjof Schuon, an outstanding poet, painter and metaphysician, in the sense that in the totality of the writings of these three metaphysicians traditional metaphysics is now being presented in all its depth and amplitude.[22]

What we are mainly concerned with here now is this question: to what extent can we identify the body of metaphysical criticisms of evolution with this general body of traditional teachings itself? We have identified earlier the origin of these metaphysical criticisms, historically speaking, with the first true revival of traditional teachings in the West associated with the above three names. Each of them did, in fact, criticize the theory of evolution on various occasions in the process of expounding their metaphysical doctrines. René Guénon, for example, criticized evolution in his exposition of the traditional doctrine of the hierarchy of existence or the multiple states of being[23] and the theory of cosmic cycles[24] among others; Coomaraswamy discussed in several of his essays the distinction between the traditional doctrine of gradation and the modern theory of evolution;[25] as for Frithjof Schuon, his reference to and criticisms of evolution were made during discussions of such doctrines as creative or cosmogonic emanation, which is an aspect of the Principle-Manifestation relationship.[26] In all these criticisms, the fundamental ideas associated with the creationism of the nineteenth century namely the immutability of species, divine archetypes, creation and design in Nature, which were described by

---

21 Nasr, S. H., *op. cit.*, p. 105.

22 *Ibid*, p. 107.

23 See his "Oriental Metaphysics" in Needleman, J., (ed), *op. cit.*, pp. 40-56.

24 René Guénon, *op. cit.*, p. 50.

25 Coomaraswamy, A. K., *op. cit.*, Chapter VI, pp. 118-124. See also his *Time and Eternity*, pp. 19-20.

26 See his *Form and Substance in the Religions*, (World Wisdom, Bloomington, Indiana, 2002), pp. 63-65: and also his *Stations of Wisdom*, (World Wisdom Books, Bloomington, Indiana, 1995) pp. 93-95.

evolutionists as negative statements about the origin and diversity of life devoid of any scientific meaning, were elaborated in detail from the metaphysical points of view. These metaphysical explanations provide the true basis for any alternative biological theory to evolution.

Having discussed and identified the origin of metaphysical criticisms we now look at their development. We need to explain here what we mean by the development of metaphysical criticisms of evolution. In a sense we can speak of traditional metaphysics as a whole as an implied criticism of evolution and all its generalizations and implications inasmuch as metaphysics is a *theoria* or vision of Reality and evolutionism is its modern substitute. That is to say, all metaphysical criticisms that there can be are contained, potentially speaking, in this general body of traditional metaphysics which has now been made available in its fullness in the language of contemporary scholarship. But there remains the work of scholarship to identify these "potential" criticisms with concrete aspects and situations pertaining to evolution and its implied world-view. It is in this area that we can speak of the development of metaphysical criticisms.

There is one more sense in which we can speak of the development of such criticisms. Once a particular individual has formulated and developed a particular criticism based on the relevant metaphysical doctrines, how is this criticism received and what is its circle of influence within the scholarly world? Development in the former sense is "vertical" and "qualitative." It refers to ideas as such irrespective of the numerical strength of its believers. It is possible that the ideas in question are subscribed to by one individual alone and then opposed or rejected by the whole academic community. However, as it stands today, there are a number of contemporary scholars belonging to the traditional world-view who have developed further the metaphysical criticisms of evolution contained in the pioneering works of René Guénon, Ananda Coomaraswamy and Frithjof Schuon. Among them we can mention Titus Burckhardt, Martin Lings and Seyyed Hossein Nasr.[27] As for the development of metaphysical criticisms in the second sense, it is

---

27 Burckhardt's detailed criticisms of evolution can be found in his "Cosmology and Modern Science" [*Editor's note:* Included in the current anthology]; For Martin

"horizontal" and quantitative. It refers to the extent of diffusion and dissemination of criticisms formulated by the above traditional scholars within the academic community. This, no doubt, depends much on the extent of influence of traditional metaphysics itself for these metaphysical criticisms can hardly be appreciated without a prior appreciation of the latter. This is best illustrated by the fact that the scholars who have dealt with metaphysical criticisms of evolutionary theory are those who have been attracted to or influenced by the traditional teachings, wholly or partially.[28]

As for the influence of traditional metaphysics in contemporary scholarship, Professor Nasr has presented us with the following assessment:

> The traditional point of view expounded with such rigor, depth and grandeur by René Guénon, Ananda Coomaraswamy, and Frithjof Schuon has been singularly neglected in academic circles and limited in diffusion as far as its "horizontal" and quantitative dissemination is concerned. But its appeal in depth and quality has been immeasurable. Being the total truth, it has penetrated into the hearts, minds, and souls of certain individuals in such a way as to transform their total existence. Moreover, ideas emanating from this quarter have had an appeal to an even larger circle than that of those who have adopted totally and completely the traditional point of view, and many scholars and thinkers of note have espoused certain basic traditional theses.[29]

We end our discussion of metaphysical criticisms of evolution with a look at their content itself. It is not possible to present here all the metaphysical arguments which have been brought against the theory of evolution. For a more complete account of these arguments we refer to the relevant works of various traditional authors that we have cited. Here we restrict ourselves to the criticisms of what we consider to be the fundamental ideas of evolutionary

Lings's criticisms, see his "Signs of the Times" in the Needleman, J., *op. cit*, pp. 109-121 and *Ancient Beliefs and Modern Superstitions*, Unwin Paperbacks, London (1980); as for Nasr's criticisms see in particular his *Man and Nature*, pp. 124-129, *Islam and the Plight of Modern Man*, Longman, London (1975), pp. 138-140 and *Knowledge and the Sacred*, pp. 234-245.

28 One can mention among them Huston Smith with his *Forgotten Truth: The Primordial Tradition*, Harper and Row, New York (1976), Chapter 6; E. F. Schumacher with his *Guide for the Perplexed* and Richard L. Thompson with his *Mechanistic and Nonmechanistic Science*.

29 Nasr. S. H., *Knowledge and the Sacred*, p. 109.

theory. In any theory, there is none more fundamental than the very basis of its own existence. And metaphysics criticizes evolutionary theory at its very root. This means that no amount of facts accumulated by biology can in any way affect the truth of this metaphysical criticism. Frithjof Schuon expressed this criticism as follows:

> . . . what invalidates modern interpretations of the world and of man at their very root and robs them of every possibility of being valid, is their monotonous and besetting ignorance of the supra-sensible degrees of Reality, or of the "five Divine Presences.". . . For example, evolutionism—that most typical of all the products of the modern spirit—is no more than a sort of substitute: it is a compensation "on a plane surface" for the missing dimensions. Because one no longer admits, or wishes to admit, the supra-sensible dimensions proceeding from the outward to the inward through the "igneous" and "luminous" states to the Divine Center, one seeks the solution to the cosmogonic problem on the sensory plane and one replaces true causes with imaginary ones which in appearance at least, conform with the possibilities of the corporeal world. In the place of the hierarchy of invisible worlds, and in the place of creative emanation—which it may be said, is not opposed to the theological idea of the creatio ex nihilo, but in fact explains its meaning—one puts evolution and the transformation of species, and with them inevitably the idea of human progress, the only possible answer to satisfy the materialists' need of causality.[30]

From the point of view of metaphysics then the true cause or origin of life does not reside in the material or physical world but in the transcendental. Objects in the world "emerge" from what is called in Islamic metaphysics the "treasury of the Unseen" (*khazânay-i ghayb*). Nothing whatsoever can appear on the plane of physical reality without having its transcendent cause and the root of its being *in divinis*. How does life "emerge" from this "treasury of the Unseen" into the physical world? This process of "emergence" can best be explained by the doctrine of the "five Divine Presences" to which Frithjof Schuon referred. The various degrees of reality contained in the Divine Principle are in ascending order, the following: *firstly*, the material state (gross, corporeal and sensorial); *secondly*, the subtle (or animistic) state; *thirdly*, the angelic world

---

30 Frithjof Schuon, *Form and Substance in the Religions*, Bloomington, Indiana (2002), pp. 63-65.

(paradisiac, or formless or supra formal); *fourthly*, Being (the "qual-ified," "self-determined" and ontological Principle); and *fifthly*, Non-Being or Beyond-Being (the "non-qualified" and "non-deter-mined" Principle which represents the "Pure Absolute").[31]

Now the formal world—the corporeal and subtle states—pos-sesses the property of "congealing" spiritual substances, of individu-alizing them and at the same time separating them one from another. Let us apply this property of the formal world to explain the appearance of species in the physical world. A species is an "idea" in the Divine Mind with all its possibilities. It is not an indi-vidual reality but an archetype, and as such it lies beyond limitations and beyond change. It is first manifested as individuals belonging to it in the subtle state where each individual reality is constituted by the conjunction of a "form" and a subtle "proto-matter," this "form" referring to the association of qualities of the species which is there-fore the trace of its immutable essence.[32]

This means that different types of animals, for example, preex-isted at the level immediately above the corporeal world as non-spatial forms but clothed with a certain "matter" which is of the subtle world.[33] These forms "descended" into the material world, wherever the latter was ready to receive them, and this "descent" had the nature of a sudden coagulation and hence also the nature of a limitation or fragmentation of the original subtle form. Thus species appear on the plane of physical reality by successive "mani-festations" or "materializations" starting from the subtle state. This then is the "vertical" genesis of species of traditional metaphysics as opposed to the "horizontal" genesis of species from a single cell of modern biology.

In the light of the above metaphysical conception of the origin of species, it is safe to say that those "missing links" which are so much sought after by evolutionists in the hope of finding the ances-tors of a species will never be found. For the process of "material-ization" going from subtle to corporeal had to be reflected within the material or corporeal state itself so that the first generations of

---

31 *Ibid*, p. 142.
32 Titus Burckhardt, *"Traditional Cosmology and Modern Science,"* op. cit.
33 *Ibid*, p. 148.

a new species did not leave a mark on the physical plane of reality.[34] It is also clear why a species could not evolve and become transformed into another species. Each species is an independent reality qualitatively different from another; this reality can in no way be affected by its history on the corporeal domain. However, there are variations within a particular species and these represent diverse "projections" of a single essential form from which they will never become detached; they are the actualization of possibilities which had preexisted in the archetypal world and this is the only sense in which we can speak of the growth and development of species.[35] In this connection, Douglas Dewar, an American biologist who was an evolutionist in his youth but later became a critic of the evolutionary theory, remarked that the whole thesis of the evolution of species rests on a confusion between species and simple variation.[36]

Metaphysics has also something to say about those biological "facts" such as the existence of "imitative" animal forms and the successive appearance of animal forms according to an ascending hierarchy which have been cited by evolutionists as clear proofs of their theory as well as the implausibility of the immutability of species. For a discussion of the metaphysical significance of these biological facts we refer to Burckhardt's essay in this anthology. We conclude our discussion of metaphysical criticisms of evolutionary theory with the following assertion: Traditional metaphysics is fully qualified to provide a meaningful interpretation to both the accomplished facts of evolutionary biology and its outstanding difficulties.

Scientific Criticisms

We now turn to a discussion of scientific criticisms of evolution, the only kind of criticisms which matter to most people today, particularly the scientific community.[37] There is as yet no complete account of the history of scientific opposition to the theory of evo-

---

34 *Ibid*, pp. 148-149.

35 Nasr. S. H., *op. cit.*, p. 235.

36 Douglas Dewar, *The Transformist Illusion*, Murfreesboro, Tennessee., Dehoff Publications, (1957). Quoted by Burckhardt, *op. cit.*

37 ". . . the only objections to evolutionary theory about which the scientists care are the truly scientific ones. These real scientific objections were the actual basis for

lution. There have been, however, several studies devoted to nine-teenth-century criticisms of evolution by the scientific community both before and after the publication of Darwin's *The Origin of Species*.[38] Studies on pre-*Origin* criticisms were carried out more with the aim of identifying the forerunners of Darwin than of under-standing the nature and dynamics of the criticisms as such. As for twentieth century scientific opposition, very little attention has been paid to it by historians and philosophers of science. There are no available sources on both the quantitative and qualitative extent of scientific criticisms of evolution in this century except for the few, but highly useful, writings of those traditional scholars we have previously mentioned.[39] We may also mention such works as Douglas Dewar's *The Transformist Illusion*, E. V. Shute's *Flaws in the Theory of Evolution* and W. R. Thompson's essay which appeared as an introduction to Everyman's Library's 1958 edition of Darwin's *The Origin of Species* replacing that of the famous English evolu-tionist, Sir Arthur Keith.[40]

From the above few works, particularly the last three, we never-theless have highly valuable information about the status of the theory of evolution within the scientific community, especially during the first half of this century. Among the important conclu-sions which can be drawn from them are: *first*, throughout its history, the theory of evolution has been continuously criticized or opposed by a section of the scientific community; *secondly*, evolu-tionists resorted to various unscientific practices in their over-zealous attempts to ensure the dominance and supremacy of

the convening of the symposium. The burden of them all was that there are miss-ing factors in present day evolutionary theory." Peter Medawar's concluding remarks as chairman of a symposium already mentioned. Quoted by A. E. Wilder-Smith in his *The Creation of Life*, p. 38.

38 See for example Gillespie, Neal C., *op. cit.*; David L. Hull, *Darwin and His Critics: The Reception of Darwin's Theory of Evolution by the Scientific Community*, Harvard University Press, Cambridge (1973); Sir A. Keith, *Darwinism and its Critics*, (1935) and the already cited *Forerunners of Darwin*.

39 *Editor's note:* Since the first edition of this article there have been a number of interesting works in this domain. For an account of these resources Chapter 7 of the current anthology by James S. Cutsinger is a very good source.

40 W. R. Thompson, "The Origin of Species: A Scientist's Criticism" in *Critique of Evolutionary Theory*, ed. Bakar, O. The Islamic Academy of Science (ASASI) and Nurin Enterprise, Kuala Lampur, Malaysia (1987), pp. 15-39.

evolutionary theory not only within the scientific establishment but also among the public at large; *thirdly*, at the beginning of the second half of the century we can detect a significant increase in the volume of scientific criticisms against various aspects of evolutionary theory of which the above three works are the best examples, and this trend has continued ever since; and *fourthly*, many scientists have expressed doubt about the general usefulness of evolutionary theory to the whole discipline of biological sciences. We will discuss these four points following our brief treatment of the issue of scientific opposition to evolution in the nineteenth century.

What we mean by scientific criticism or opposition here is that the nature of the arguments is scientific as this term is generally understood today, rather than that the source of the arguments is scientific. In the nineteenth-century debate on evolution, this distinction has to be made because there were many scientists who opposed the new theory on both scientific and religious grounds. These include, at least until the publication of the *Origin*, such well-known scientists as the American geologist Edward Hitchcock, British geologist Adam Sedgwick, Richard Owen,[41] England's foremost comparative anatomist in the 1850s, Louis Agassiz and James Dwight Dana, the two most influential of American naturalists, geologist Joseph LeConte who was Agassiz's student, the English entomologist T. Vernon Wollaston, Scottish naturalist the Duke of Argyll, Canadian scientist John William Dawson, mathematician-geologist William Hopkins and many others.[42] All of them rejected evolution then as contrary to known geological and biological facts.

Not long after *The Origin*, many scientists were converted to the evolutionary doctrine including a former critic Joseph LeConte mentioned above. Others like Richard Owen, the Duke of Argyll and St. George Jackson Mivart who published his *Genesis of Species* in 1871 adopted an intellectual compromise between their former position and Darwinian evolution through their idea of providential evolution. In reality, however, the two kinds of evolution do not differ in intellectual substance or doctrinal content for they refer to the same organic process.[43] Where they differ is in their views of the

---

41 On their critiques see Gillespie, N. C., *op. cit.*, p. 22.
42 *Ibid*, p. 26.
43 *Ibid*, Chapter 5, entitled "Providential Evolution and the Problem of Design."

place and role of God in that process. For the Darwinian evolutionists, organic evolution is purely a product of physical and natural
causes while for the providential evolutionists it is God's mode of
creation. Though the providential evolutionists vehemently
opposed Darwin's natural selection as an explanatory mechanism
of organic evolution insofar as it leaves no room for divine purpose
and control, their acceptance of organic evolution albeit in religious shape "with little touches of special creation thrown in here
and there"[44] took them closer to positivism and out of the realm of
special creation. As for the rest of the scientists like Louis Agassiz
who believed in special creation and continued to oppose the idea
of evolution, they became a rarer intellectual species by the end of
the century though by no means extinct.

In the light of oft-repeated charges that the theory of evolution
has no scientific basis whatsoever, we should investigate what then
caused the conversion of a large number of scientists to the evolutionary doctrine after the publication of *The Origin*. Certainly it was
not due to the convincing amount of scientific evidence marshaled
by *The Origin*. On the contrary, Darwin himself referred more than
once to the lack of evidence in support of many of his claims in *The
Origin*. The success of the theory of evolution was due mainly to
factors other than scientific. In fact we can assert categorically that
there was something very unscientific about the whole way in which
the theory rose to its dominant position in science, and as we shall
see later, also about the way in which it has attempted to maintain
this dominance. It became dominant not through its own strength
by which it withstood tests, analyses and criticisms but through the
weakness of its rivals, those various forms of creationism which were
in conflict with each other and which no longer satisfied the positivist's need for causality. Since the theory is a fruit of the application of the philosophical idea of progress to the domain of biology,
the ascendancy of the latter idea in the nineteenth century contributed greatly to the ascendancy of the theory. Thus it has been
said:

> . . . the theories of evolution and progress may be likened to the
> two cards that are placed leaning one against the other at the foun-
> dation of a card house. If they did not support each other, both

---

44 *Ibid*, p. 103.

would fall flat, and the whole edifice, that is, the outlook that dominates the modern world, would collapse. The idea of evolution would have been accepted neither by scientists nor by "laymen" if the nineteenth-century European had not been convinced of progress, while in this century evolutionism has served as a guarantee of progress in the face of all appearances to the contrary.[45]

There was no lack of scientific arguments on the part of nineteenth-century critics of evolution. But somehow the evolutionists did not address themselves fully to the fundamental issues and objections raised in these scientific arguments but instead highlighted the inadequacy and negativity of creationism as explanatory mechanisms of the diversity of living organisms.

Let us return to the "four points" previously mentioned. First, we said that the theory of evolution has been continuously opposed by a section of the scientific community. From the 1890s to the 1930s there was a widespread rejection of natural selection among the scientific community.[46] Though the rejection of natural selection does not necessarily imply the rejection of evolution itself, it does show that the true explanation of biological diversity has not yet been found and without any plausible mechanism of how evolution has occurred the status of evolution is nothing more than that of a hypothesis at best. In their continuing efforts to defend the idea of evolution, numerous explanations were offered by various scientists as to how it has occurred but in the words of Dewar they were all purely conjectural and mutually contradictory.[47] There is also the admission by a Sorbonne Professor of Paleontology, Jean Piveteau, that the science of facts as regards evolution cannot accept any of the different theories which seek to explain evolution and in fact it finds itself in opposition with each one of these theories.[48]

The general disagreement among scientists on this very question continue until this very day. Only very recently, this internal controversy within the evolutionary ranks became a near battle when some 150 prominent evolutionists gathered at Chicago's Field Museum of Natural History to thrash out various conflicting hypotheses about the nature of evolution. After four days of heated

---

45 Martin Lings, *"Signs of the Times,"* in Needleman, J., *op. cit.*, p. 112.
46 Gillespie, N. C., *op. cit.*, p. 147.
47 Martin Lings, *Ancient Beliefs and Modern Superstitions*, pp. 5-6.
48 *Ibid*, p. 5.

discussions (closed to all but a few outside observers), the evolutionists remained convinced that evolution is a fact. In reality, this was an affirmation of faith rather than of fact because, as *The New York Times* reported it, the assembled scientists were unable either to specify the mechanisms of evolution or to agree on "how anyone could establish with some certainty that it happened one way and not another."[49] One of the participants, Niles Eldredge, a paleontologist from the American Museum of Natural History in New York, declared: "The pattern we were told to find for the last 120 years does not exist."[50]

The above conflict and confusion among evolutionists only serves to confirm the belief of many critics of evolution that that is what is bound to happen once scientists start looking at the theory critically. This brings us to our second and third points. The increase in the volume of scientific criticisms in the beginning of the second half of this century can partly be attributed to a certain level of tolerance toward criticisms, in comparison to the earlier decades, as attested by the replacement of Arthur Keith's evolutionary hymn in the introduction to *The Origin* by Thompson's critical introduction in 1959. It also coincided with the beginning of skepticism of "progress" itself in the aftermath of the Second World War. As for the first half of the century, it was a period of unquestioned faith in evolution,[51] intellectual intolerance and dishonesty on the part of many evolutionists. Intellectual intolerance and dishonesty manifests themselves in many ways. For example, there are cases of intolerance in the form of opposition against those types of research work which seek to explain biological phenomena in non-evolutionary terms. One such case was the attempt of D'Arcy Thompson to explain embryological development in terms of actual physical causes rather than to be content with explanations of a phylogenetic nature, but this was rejected with contempt by authors like Haeckel and other evolutionists.[52] As for intellectual dishonesty, one may refer to the famous hoax connected with the

---

49 Richard L. Thompson, *op. cit.*, pp. 183-184.
50 *Ibid*, p. 185.
51 Thompson, F. R. S., *Science and Common Sense*, London (1937), p. 229.
52 W. R. Thompson, *op. cit.*, pp. 15-39.

alteration of the Piltdown skull so that it could be used as evidence for the descent of man from the apes.

On the question of usefulness of evolutionary theory to biology, many biologists have expressed the opinion that the latter would have achieved far greater progress had it not been addicted to evolutionary thinking. They do not dispute the fact that evolution has greatly stimulated biological research, but owing precisely to the nature of the stimulus a great deal of this work has been directed into unprofitable channels. Too much time, labor and scientific talent were wasted in the production of unverifiable family trees, the tracing of ancestries or the construction of hypothetical ancestors and unverifiable speculations on the origin of structures, instincts and mental aptitudes of all kinds. To the point raised by evolutionists that a vast amount of biological facts has been gathered in these studies, these critics express the belief that they could have been obtained more effectively on a purely objective basis.[53]

Scientific criticisms of evolution do not come from biologists only. There is also an increasing number of scientists in other disciplines, particularly physicists and mathematicians, who have criticized the theory of evolution from the viewpoint of present knowledge in their respective fields. Richard L. Thompson, an American mathematician who specialized in probability theory and statistical mechanics and who has done research in mathematical biology, has argued in his *Mechanistic and Non-mechanistic Science: An Investigation into the Nature of Consciousness and Form* that the theory of evolution is not actually supported by the factual evidence of biology and natural history. Drawing on ideas from information theory, Thompson shows that configurations of high information content cannot arise with substantial probabilities in models defined by mathematical expressions of low information content.[54] This means that complex living organisms, which possess a high information content, could not arise by the action of physical-chemical laws considered in modern science, since these laws are represented by mathematical models of low information content. Thompson defines the information content of a theory to be "the

53 *Ibid.*
54 See Thompson, R. L., *op. cit.*, p. 97.

length of the shortest computer program that can numerically solve the equations of motion for the theory to within any desired degree of accuracy."[55] His fundamental argument is that in a physical system governed by simple laws, any information present in the system after transformations corresponding to the passage of time must have been built into the system in the first place. Random events cannot give rise to definite information, even when processed over long periods of time according to simple laws. On the basis of these fundamental arguments in information theory, Thompson maintains that the existence of a complex order here and now cannot be explained unless we postulate the prior existence of an equivalent complex order or that the information content of the system has been received from an outside source.

The consequence for the idea of organic evolution is clear. The process of natural selection, accepted by many scientists as the mechanism of evolution, could not have brought about the development of complex living organisms because the laws of nature (currently conceived) underlying the process lack the necessary information content to specify its direction.

There are other scientists who with the aid of information theory have arrived at a similar conclusion concerning the current theory of evolution. The eminent British astronomer Sir Fred Hoyle and the distinguished astrophysicist Chandra Wickramasinghe, both of whom were once agnostics, draw the following conclusion from their study of recently assembled facts in such disciplines as microbiology, geology and computer technology: the complexity of terrestrial life cannot have been caused by a sequence of random events but must have come from some greater cosmic intelligence.[56]

It is not possible within the scope of this essay to go into the detailed scientific criticisms that have been put forward up till now against the evolutionary theory. The main message we seek to convey is that scientific opposition against evolution is gaining momentum. These scientific criticisms, coming as they are from different sciences, call into question the status of evolutionary doc-

---

55 *Ibid*, p. 105.
56 See Hoyle, Sir Fred and Wickramasinghe, N. C., *Evolution from Space. A Theory of Cosmic Creationism*, New York (1981).

trine as the integrative principle of all the sciences which is being claimed by many evolutionists.

Religious and Philosophical Criticisms

Besides scientific and metaphysical criticisms, there are the religious and philosophical ones. From the religious points of view the evidence against evolution is universal. In all sacred Scriptures and traditional sources whether they speak of creation in six days or of cosmic cycles lasting over vast expanses of time, there is not one indication that higher life forms evolved from lower ones. Says Professor Nasr: "The remarkable unanimity of sacred texts belonging to all kinds of peoples and climes surely says something about the nature of man."[57] As for philosophical criticisms, Thompson referred to the opinion of respectable philosophers who hold that the Darwinian doctrine of evolution involves serious difficulties which Darwin and others like Huxley were unable to appreciate. They argued that between the organism that simply lives, the organism that lives and feels, and the organism that lives, feels and reasons, there are abrupt transitions corresponding to an ascent in the scale of being and that the agencies of the material world cannot produce transitions of this kind.[58] Philosophers such as Michael Polanyi and Karl Popper have criticized the current theory of evolution, though their philosophical alternative is unacceptable from the view point of metaphysics. Says Polanyi:

> Scientific obscurantism has pervaded our culture and now distorts even science itself by imposing on it false ideals of exactitude. Whenever they speak of organs and their functions in the organism, biologists are haunted by the ghost of "teleology." They try to exorcise such conceptions by affirming that eventually all of them will be reduced to physics and chemistry. The fact that such a suggestion is meaningless does not worry them . . . the shadow of these absurdities lies deep on the current theory of evolution by natural selection.[59]

57 Nasr, S. H., *op. cit.*, p. 237.
58 W. R. Thompson, *op. cit.*, pp. 15-39.
59 Polanyi, M., *Knowing and Being*, University of Chicago Press, Chicago (1969), p. 42.

Conclusion

What do all these criticisms, metaphysical, scientific, religious and philosophical, mean to the future of the theory of evolution? We have no doubt that if the theory is allowed to be scrutinized critically and openly by all interested parties the collapse of evolutionary theory is in sight. The skepticism that is now current of the idea of progress will also have a great impact on the future of evolution since it has been the very basis of its origin, ascendancy and survival. Anyway there are already those who are very definite about what is going to happen to the theory. Says Tom Bethell:

> Darwin's theory, I believe, is on the verge of collapse. . . . He is in the process of being discarded . . .[60]

---

60 Quoted by Huston Smith, *Forgotten Truth: The Primordial Tradition*, Harper and Row, New York (1977), p. 134.

9

Knowledge and Knowledge

D. M. Matheson

Man is born with a thirst for knowledge of one kind or another. And in the sphere of science and technology with which the Western world is now so preoccupied each generation adds to the mass of accumulated data, which thus mounts in geometrical progression, doubling every fourteen years or so. Now, because in that sphere there is a need for a particular kind of precision, there has been a tendency to look on the form of logic which says that "A is not both A and not-A" and that "A is either B or not-B" as the highest form of thinking for all sorts of purposes though it often leads to what are in fact correlatives being envisaged as antagonistic opposites. Indeed it is inadequate for many scientific purposes as, for example, when the dual wave and particle aspects of the electron are being considered, moreover ecologists are obliged in some degree to share the view that "the universe is a system in which every element, being correlative to every other, at once presupposes, and is presupposed by every other."

One example of this type of logic is that we talk about man as an animal and suppose that he cannot also be not-an-animal, and we are fortified in this view by a widely held view of the origin of life and of consciousness.

In a famous lecture in 1874 Tyndall asserted that "in matter lies the promise and potency of every form and quality of life." And, whereas Plotinus had held on metaphysical grounds that "the idea that elements devoid of intelligence should produce intelligence is most irrational," Bertrand Russell assures us that "man's origin, his growth, his hopes and fears, his loves and beliefs are but the outcome of accidental collocations of atoms." And, whereas Pasteur's experiments were at one time thought to have established the dictum *omne vivum ex vivo*, today we are told that over some thousands of millions of years the blind working of physical forces has accidentally led from atoms to molecules, from molecules to living cells and so to man, to his consciousness and reason. Presumably the vast time interval makes the theory sound more rationally acceptable. But we are also told that, apart from an infinitesimal element of indeterminacy, our thoughts, feelings and actions are all determined by inherited, ultra-microscopic, physico-chemical genes or by the interaction between the organism and environmental forces equally physico-chemical in origin.[1]

All this does not prevent those who hold such views from behaving as if they believed their thoughts and actions to be determined by their own free will; often, indeed, they say that by conscious use of the resources of science man can indefinitely perfect—by what standards?—both man himself and his circumstances!

Let us remind ourselves that scientific observation does not see the world as it is in itself; there is always an element of the subjec-

---

1 *Editor's note:* "In view of the fact that modern science is unaware of the degrees of reality, it is consequently null and inoperative as regards everything that can be explained only through these degrees, whether it be magic or spirituality, or any belief or practices of any peoples; and it is in particular incapable of accounting for human or other phenomena situated in a historical or pre-historical past whose nature, or key, eludes it totally or as a matter of principle. Thus there can be no more desperately vain an illusion—far more naive than is Aristotelian astronomy—than to believe that modern science will end up reaching, through its dizzying course towards the 'infinitely small' and the 'infinitely great,' the truths of religious and metaphysical doctrines.... Of all this, experimental and pragmatic science knows nothing; the unanimous and millenary intuition of human intelligence means nothing to it, and scientists are obviously not prepared to admit that, if myths and dogmas are so diverse, in spite of their agreement with regard to the essential—namely, a transcendent and absolute Reality and, for man, a hereafter conforming to his terrestrial attitudes—this is because the supra-sensory is unimaginable and indescribable and allows for indefinitely varied ways of seeing adapted to different spiritual needs. The Truth is one, but Mercy is diverse. Not only is scientistic philosophy ignorant of the Divine Presences; it ignores their rhythms and their 'life': it ignores, not only the degrees of reality and the fact of our imprisonment in the sensory world, but also cycles, the universal *solve et coagula*; this means that it ignores the gushing forth of our world from an invisible and fulgurating Reality, and its re-absorption into the dark light of this same Reality. All of the Real lies in the Invisible; it is this above all that must be felt or understood before one can speak of knowledge and effectiveness. But this will not be understood, and the human world will continue inexorably on its course." Frithjof Schuon, *Form and Substance in the Religions,* (World Wisdom, Bloomington, Indiana, 2002) p. 65.

tive and anthropomorphic, always a chasm between language and reality. The popular illusion that physics has now understood and explained the real nature of the world, of the whole of manifestation, is by no means always shared by the physicists. "Leaving out," said Eddington, "all aesthetic, ethical or spiritual aspects of our environment, we are faced with qualities such as massiveness, substantiality, extension, duration, which are supposed to belong to the domain of physics. In a sense they do belong; but physics is not in a position to handle them directly. The essence of their nature is inscrutable; we may use mental pictures to aid calculations, but no image in the mind can be a replica of what is not in the mind. And so in its actual procedure physics studies not these inscrutable qualities but pointer readings which we can observe. The readings, it is true, reflect the fluctuations of the world-qualities: but our exact knowledge is of the readings, not of the qualities. The former have as much resemblance to the latter as a telephone number has to a subscriber."

Of course man is an animal and as such motivated by an animal will to live and to breed in the fierce competitive struggle common to all forms of life. Moreover, as a social animal he is also conditioned by the will of the group to survive and prosper in competition with other groups; this implies that conformity to the law or needs of the group must be enforced and any nonconformity dangerous to the group must be punished. Marxist societies, which feel themselves to be surrounded by hostile communities, have dealt ruthlessly with any deviation and only effectively conditioned individuals are allowed to remain long in close contact with ideologies they regard as poisonous; in Western democracies this consequence of man's status as a social animal tends to be slurred over, often sentimentally.

What has almost vanished today in Europe and America is the idea formerly current that besides the ordinary particulate and accumulating knowledge in which, through our schooling, we are all in some degree partakers, there is also another kind of knowledge, a knowledge imparticulate and incommensurable with our ordinary knowledge. Of this knowledge there could be no quantitative accumulation as in the case of technical and scientific data; it was held indeed to be indescribable and in a sense incommunicable since it was associated with a different state of being characteristic of sages, seers and saints. Traces of this idea can be seen in the distinction made by the Greeks between *noesis* and *dianoia* and in the mediaeval scholastic use of the term "intellect." Implicit in it is the idea that man is not only an animal but also not-an-animal.

Heraclitus pointed out that the end of strife—of the contraries—would mean the destruction of the universe but that men "fail to grasp that what is at variance agrees with itself in an attunement of opposite tensions as in the bow or the lyre." With striking unanimity all the great religious traditions indicate that on the scale of a man there is, at any rate in certain cases, a possibility of transcending duality—or the contraries—in this life and coming to a new, supra-human state of being and a new kind of knowledge and that this possibility implies also a destruction. If, as has been said, this new kind of knowledge is indescribable, its nature has nonetheless been indicated through the use of paradox and symbols and its quality has been described as Bliss.

As animal man is at least in large measure conditioned by his environment, and, if the environment is chaotic and full of contradictions, its chaos will be reflected in him. It is the claim of the great traditions that they have provided an environment, supernatural in origin, which is a reflection of objective truth and thus free from inner contradictions and full of symbolisms. Such an environment they would claim to be a prerequisite for any supernatural change in man giving access to this second kind of knowledge.

Admittedly, once influences supernatural or divine in origin or inspiration become embedded in forms those forms come under the laws of decay and mortality imposed by devouring time on all forms and organisms, and it is all too easy to point out evidences of this in traditional forms known to us. Indeed, a "materialistic" modern outlook could not otherwise have gained such a fascinated acceptance. There has been a degree of failure on the part of Christian leaders to offer a picture of man and the world in their total setting adequate to satisfy intellectual needs, and in the resulting void man—ordinary "animal" man—has been enthroned in the place of God, and religion has often evaporated into morality and humanitarianism. The very idea, characteristic of traditional esotericism, of a possibility of deliverance "here and now" into a different knowledge and being has all but vanished.

The Great Traditions

But the great traditions have not wholly fallen under the law of decay. In all of them seers, prophets, sages and saints have actualized this different knowledge and being and have thus represented fresh influxes of divine influence to revivify the traditions from which they sprang. And, if the popular tales of their lives are often richly embellished with miraculous manifestations, this is at least in part a symbolical or poetic expression of the fact that they were themselves a miracle. Whereas we are conditioned by, or slaves to, a thousand influences from our passions and our environment they are delivered from such slavery into a new, supernatural kind of knowledge and that service which is perfect freedom, and it is not surprising if such freedom has at times found expression in ways highly shocking to formalist "doctors of the law" of their tradition.

One side effect of the feeling of an intellectual void in our society has been the growth of interest in and study of those Oriental doctrines and disciplines which are said to lead to new knowledge and a different state of being. The trouble is that such studies are almost inevitably limited to certain fragments divorced from the total traditional framework which should normally condition the whole psychic background. In a Hindu world, for instance, the whole of life is interwoven with traditional art, myth and ritual rich in symbolism capable of conveying aspects of truth which books and mental studies cannot impart, and the direct personal help and guidance of a Hindu master presupposes all these elements having played their part. It is not the thinking mind which balances the body, which falls in love or discriminates between "me" and "other-than-me," nor is it by a mental process that such "horizontal" discrimination can be transcended through qualitative discrimination between different levels of manifestation. Nor is it by will power that the axe of discrimination can be wielded.

Anyone who seeks to find his way to this second kind of knowledge must get free from three knots in the bonds which bind him. The first knot is that, whether we admit it or not, we very often identify "me" with the body. The second is that we are under the domination of desires which we also identify with "me"; and it should be noted that the *apatheia* spoken of by Christian Fathers means, not "apathy" but an active control which liberates from this domination. Thirdly, we identify the workings of the mind with "me" —and this knot is by far the most difficult to unravel. What am I if not my mind? The answer can really only be discovered through experience and one difficulty is that "I" cannot loose these knots; it requires the power of "other-than-I" and the further question arises: who is this "other-than-I"? And it is precisely the traditional forms, rites and symbolisms, which we are inclined to discount as mere exoteric formalism, that can help us to answer this question.

We are apt to envisage the process of coming to a new kind of knowledge as the acquiring of new powers and increased efficiency. It is true that in the preparatory process there must be a change in our center of gravity, a reduction of inner chaos and a new harmony in our ideas which may incidentally yield such results and also enlarge our field of vision, but to come to such knowledge means much more than this; it involves something exceedingly painful to "me"—extinction of the ego and of the sense of separateness. In the deepest sense there is nothing to be acquired.

Some people who hear of these possibilities doubt if they really exist. On the basis of the modern quantitative and egalitarian outlook they ask why, if they exist, they seem to be so very rarely actualized. One doesn't meet such men, they say. Let us recall the story of how a sage who saw the infant prince Gautama foretold that he would be either a Buddha or else a world-conqueror. Even among those who feel a call to seek such knowledge through appropriate means potential world-conquerors are rare indeed! Some are easily bewildered and led astray, many are relatively feeble. "Knock and it shall be opened to you," said Christ, but he added that the way to Life is narrow and found by few. To knock successfully at the door leading to the second kind of knowledge involves finding the right door at which to knock and then knocking both with great persistence and with that skill in action which is one of the definitions of Yoga; nor must we leave out of account what is called in religious terms Divine Grace.

All this sounds very discouraging of any aspirations to such knowledge since it is obvious that of the few who set out on a path to it very many are likely to fail to reach the objective. But in any of the more ordinary ventures of life the really bold and determined are not easily put off by accounts of tremendous obstacles. A Hindu considering the difficulties might well say that of course many lives are needed for reaching such an objective, but Christianity and Islam do not envisage the idea of palingenesis; each tradition has its own perspective and here more emphasis is placed on the posthumous rewards of true believers. We who live with only the ordinary kind of knowledge and with all sorts of illusions about "me" cannot know about death, or about the fate of a traveler on the road to the other kind of knowledge, what is only within the ken of that knowledge; we have to go largely by faith. And all the traditions say that perseverance in a true path always brings rich rewards for those whose qualities call them to such a path.

10

Knowledge and its Counterfeits

Gai Eaton

Given the nature of the time process, it is not particularly surprising that the notion of man's viceregal function and dignity should have been forgotten. We are by nature forgetful, which is no doubt why the religion of Islam describes itself specifically as "a reminder to mankind." What is truly astonishing is that this notion should now appear nonsensical to the vast majority of people in the West and, indeed, to "educated" people everywhere. The fact that a view of man's destiny which could be considered, until so recently, as something inherent in human thinking should be dismissed as a fairy tale would be incredible if it had not actually happened.

No wonder that many of those who hold to the traditional view believe the devil himself has bewitched our kind, putting to sleep the faculties through which we were formerly aware of realities beyond the field of sense perception and making use of mirages to lead us into a waterless desert. This process culminates in a narrowing of horizons which Mircea Eliade and others have described in terms of "provincialism." We live and think and operate today within the dimensions of a wafer-thin cross-section of historical time, effectively isolated from the past as from the future.

Evolutionary theory, as it is commonly understood by nonspecialists, has penetrated very deeply into the substratum of human thought. It shapes opinion and distorts judgement in almost every sphere, all the more effectively because it has become a kind of unconscious and therefore unquestioned bias. People readily assume that each generation is likely to be a little wiser (and possibly even a little better) than the preceding one; this assumption is inherent in the idea of progress as it is commonly understood. If that were so, then the beliefs and ideas of earlier generations might reasonably be dismissed as obsolete. Religion would be no more than a vestige of primitive thought, and Christ might be considered, at best, as a man ahead of his time, a signpost on the evolutionary path. This appears to have been the view of Teilhard de Chardin, that misled and misleading priest.

We make certain deductions from the facts available to our senses in this thin slice of time. It is assumed that the people of earlier ages tried to do the same, and since they did not deduce what we have deduced from these facts they must necessarily have been our inferiors. It is taken for granted that their beliefs were based, as ours are, upon the observation of physical phenomena. They were not very good observers and persistently drew the wrong conclusions from such facts as they did observe; they belonged, it is said, to a "pre-logical" stage of human development.

This is, in the first place, a childish attitude. It is common enough for children to enjoy a sense of superiority over adults who cannot climb trees as they do or who make a mess of a jigsaw puzzle which presents no problem to an eight-year-old, and a child may reasonably wonder why a grown-up who can afford to buy ice cream or chocolates every day of his life does not do so, just as we are puzzled that the ancients never developed effective techniques for the exploitation of the earth's riches. Grown-ups, however, have a different order of priorities.

This childish aspect of modernism is nothing if not naïve in its view of the past. It takes for granted that if all we want is ice cream or its equivalents, then this is all that people ever wanted. They did not know how to produce it quickly, hygienically and in quantity. We do. They were not clever enough to invent motor cars and aeroplanes. We are (without ever asking ourselves whether our journeys are really necessary). They thought the earth was the center of the universe. We know better.

Arguments of this kind, however ludicrous they may seem, are at the root of a great deal of modern thinking, not, of course, among a sophisticated minority of scholars and intellectuals, but among ordinary people who have received the usual smattering of education and have been encouraged to believe that they know something worth knowing. What matters, from this point of view, is not the pure form of a particular theory but the form in which it has been popularized, processed through the educational machine and assimilated by the masses. Religious (or metaphysical) ideas, when they penetrate whole populations within a traditional environment, may adopt simplified and what might be described as "picturesque" forms without thereby sacrificing either integrity or effectiveness, but secular and scientific notions soon become slipshod and inaccurate when they are popularized.

Most important of all, perhaps, modern thought is "provincial" insofar as most people are confined within the narrow limits of faculties designed to deal only with our own small corner of creation and ill-adapted (as is our language itself) to anything beyond self-preservation and the getting of food. Our ideas of truth and indeed of all that is seldom go beyond the things which fit the contours of a mind as limited in its way as are our physical senses; and we are necessarily ignorant in the precise sense of the term, since it is obvious that the mind as such cannot comprehend—within its own terms of reference—what lies beyond this particular locality and the view visible from here.

The distinction between ignorance and agnosticism—a distinction which is often ignored in our time—is of great importance. The former is both natural and realistic; it knows itself and recognizes its own impotence. To be human is, in the first place, to be ignorant and to accept the fact that there is a great deal we cannot know and, for that matter, a great deal we do not need to know. Idle curiosity is certainly a vice—a lust of the mind—whereas acknowledgement of the fact that we have no intrinsic right to receive answers to all our questions is an aspect of humility as it is of realism. It is said that St. Augustine was asked: "What was God doing before he created the world?" "Preparing hell for those who ask unnecessary questions!"

Agnosticism however raises a personal incapacity to the dignity of a universal law. It amounts to the dogmatic assertion that what "I" do not know cannot be known, and it limits the very concept of what is knowable to the little area of observation open to the unsanctified and unilluminated human mentality. The agnostic attitude derives from a refusal to admit that anyone can be or ever could have been our superior in this, the most important realm of all: the true knowledge of what there is to be known. Religion is now seen exclusively in terms of faith rather than of supernatural knowledge. In egalitarian terms, faith is acceptable; you may believe in fairies if you wish to. But the claim to a direct and certain knowledge of realities beyond the mind's normal compass excludes those who do not possess it and savors of presumption. The idea that a saint among the saints may have *known* God—not merely *believed* in him—suggests "unfairness" and implies the superiority of some men to others. It puts us in our place.

Squatting in this place, this little pool, and hungry for certainties, people hold on with a kind of desperation to the current notion of what is (or is not) "rational"; and yet, "the rationalism of a frog at the bottom of a well consists in denying the existence of mountains; this is logic of a kind, perhaps, but it has nothing to do with reality."[1] This rationalism is inextricably linked with the scientific point of view, which is advanced as the only logical interpretation of the world. Unfortunately nothing in this realm is as clear as it should be. The "facts" with which science supplies us are of quite a different order to those registered by our physical senses. What the scientist says, in effect, is this: you may take my propositions as proved, provided you accept all the assumptions which appear self-evident at this time, so long as you agree that the objective world exactly fits the patterns inherent in human thinking (or vice versa), on the understanding that the simplest explanation of a given phenomenon must be the right one and assuming that the physical world is sealed off from interference from any other realm. This adds up to a formidable list of qualifications.

Contrary to popular belief, science does not offer us certainties in the way that our senses provide a kind of certainty on their own level. Scientific hypotheses are not facts, and before the scientist can even begin to construct his theories he must make a number of very sweeping assumptions which most people may agree to take for granted, since they are in accordance with the present climate of opinion, but which can never be proved.

He must assume the absolute, objective validity of his own mental processes and believe that the logic of these processes is a universal law to which everything that is or ever could be conforms. Common sense tells him that this is so, but common sense is a variable factor which changes from one age to another. He can never be certain that the images which his senses present to his mind are a true representation of realities which exist independently and objectively. Not unlike the man who interprets the outside world in terms of what is taking place in his own entrails, seeing a bright day

---

1 *Logic and Transcendence.* Frithjof Schuon, p. 42.

when he feels well and finding the world a dark place when his system is choked with waste products, he may in fact be applying to observed data the laws which govern his own mentality, an instrument constructed for the practical business of living much as the entrails are constructed for the digestion of food. Since inner and outer, subjective and objective, are, in the last analysis, two sides of the same coin, he is likely to find that the protean physical world provides the answers he expects of it (these answers being implicit in the phrasing of his questions) and experiment will confirm the theories he has constructed without, in fact, taking him beyond the subjective realm.

However complex the instruments designed to extend the range of our senses, scientific exploration is always to some extent dealing with patterns inherent in the exploring mind and meeting the mirror images it has projected. Nature mocks and eludes us, seeming to fit herself into our mental categories because our minds are themselves embedded in her structure. We imagine ourselves standing—or floating—above the natural world, competent to survey it objectively, and the intervention of scientific instruments between our naked senses and the objects of observation heightens this illusion; but a mentality which is part of the natural world can never escape and look down as a disembodied agent upon its own matrix. That element in man which does indeed transcend the natural world is in him but not of him, and the objectivity of its awareness is very different from the fictional objectivity exercised by one facet of nature in relation to another.

But while the scientist, in his increasingly private and abstract sphere, finds a marvelous concordance between his thinking process and the movement of a needle on a dial or the traces of radiation on a photographic plate, the ordinary man of our time faces a widening gulf between scientific theory and any kind of objective experience known to him.

No longer can men be told that the truth of things will be confirmed in their own intimate experience if only they will look and listen. The proofs and arguments of contemporary science are so abstract and so technical that they are no longer open to criticism by the non-specialist and cannot be tested against any kind of experience known to man as a living creature. Informed that the electron's position does not change with time, but does not remain the same, and that, although the electron is not at rest, it is not in motion, François Mauriac remarked: "What this professor says is far more incredible than what we poor Christians believe!" The theories employed by modern physics have not merely by-passed the contours of the rational mind; they have gone beyond the range of human imagination.

"In those never-never, through-the-looking-glass abodes," says Professor Huston Smith, "parallel lines meet, curves get you from star to star more quickly than do Euclid's straight lines, a particle will pass through alternative apertures simultaneously without dividing, time shrinks and expands, electrons . . . jump orbit without traversing the intervening distance, and particles fired in opposite directions, each at a speed approximating that of light, separate from each other no faster than the speed of light."[2] After this no one has any excuse for finding obscurities or improbabilities in the higher reaches of theology and metaphysics. If the majority of people still imagine that the physical sciences relate in some way to their normal experience this can only be because they are living in the past, comfortably immured in the mechanistic science of the nineteenth century.

Although in no sense supernatural, contemporary scientific theories do not relate to the spectacle of nature as we know it in our daily lives, and their "proofs" derive from experiments carried out under almost unimaginable conditions (at temperatures a fraction of a degree above absolute zero and so on) with the aid of immensely complex equipment. In factual terms—and a fact, after all, is something against which we expect to be able to stub our toes—this is a very remote and esoteric region. And it is partly because these theories, together with their proofs, are unverifiable in terms of human experience and because they originate in the extra terrestrial conditions created in the secrecy of the laboratory that they have such power to bind and fascinate. Their glassy surface offers no purchase to the mind's skeptical probing.

A field of knowledge in which the ordinary person can participate only by believing what he is told by experts corresponds very well to the political field of the totalitarian State in which he participates only by doing what he is told to do by an anonymous *them*, while the notion that every new fact discovered by science adds to

2 *Forgotten Truth*: Huston Smith (Harper & Row), pp. 105-106.

the universal store of human knowledge and that this quantitative accumulation is an unqualified good finds its echo in the belief that every technological advance contributes to the wellbeing of mankind.

Speaking of the "normal and providential limitation of the data of experience," Frithjof Schuon remarks that, while no knowledge is bad in itself and in principle, many forms of knowledge may be harmful in practice "because they do not correspond to man's hereditary habits and are imposed on him without his being spiritually prepared; the soul finds it hard to accommodate facts that nature has not offered to its experience, unless it is enlightened with metaphysical knowledge or with an impregnable sanctity." The unenlightened and unsanctified personality, subjected to a barrage of facts and theories which contradict its own intimate experience of the world, is more likely to be maimed than nourished. Through their education and by means of books, newspapers and television, people's minds are now crammed with ill-assorted fragments of knowledge. Without any unifying principle, this adds up to little more than a pile of debris which is never effectively sorted or assessed. No wonder we choke on it and lose our bearings.

People have a longing for normality or, in other words, a need to be what they are meant to be. It would be strange were this not so, but the fact remains that, when the true norm has been forgotten, it is only too easy to go off into the wilds in pursuit of a substitute. Just as nostalgia for the integral traditional society, in which everything fits and everyone has his place under the light of heaven, can draw us fatally towards the totalitarian society, so nostalgia for true and certain knowledge induces us to embrace its counterfeits and to mistake an accumulation of facts for something that they can never be. Quantity, by whatever factors it may be multiplied, is never more than a finite number, a fragment. Though you pile fact upon fact until the heap of evidence seems to touch the sky, it is still nothing in comparison with totality, just as a distance of countless light-years still comes no closer to infinity than does a single centimeter.

A counterfeit coin is still a coin, though we mistake its nature and its value. Those who are deceived may blame it, but the coin is what it is, no more, no less. Scientific knowledge is what it is, neither absolutely true nor absolutely false, but always relative and contingent. Theories based upon the observation of happenings which occur again and again in a particular cross section of time have their practical uses but can never be more than hypothetical; insofar as we take them for certainties, they are counterfeits.

To say this is not to suggest that observed facts and the general laws derived from them are without significance, but only to emphasize the fact that they belong to the realm of relativity—and therefore of uncertainty—and cannot under any circumstances emerge from this domain. They deal with phenomena in a particular theatre at a particular moment in time, but they can tell us nothing about the open, the universal, the total. They remain bound to a locality, since any given phenomenon may be "explained" in a variety of ways and at various levels; our preconceptions—and the prevailing climate of opinion—determine our choice of explanation.

At the same time, science can never allow for the ambiguity inherent in the natural world, an ambiguity which is brought out with particular clarity in the Hindu doctrine of *mâyâ*, the divine art, the divine magic, the divinely willed "illusion" which is, in a sense, all things to all men. The physical sciences deal exclusively with the slippery and deceptive realm of *mâyâ* and therefore cannot in any way determine the nature of the Absolute or, indeed, pretend to take precedence over direct, immediate knowledge on the one hand or its objective counterpart, Revelation, on the other. But what can—and does—happen is that these relativities and probabilities are inflated until they fill the view and nothing else can be seen.

Facts and the theories derived from them lodge only in the mind, whereas the metaphysical truths which lay at the root of human belief in other times transcend the personality as such and are no more exclusively mental than emotional or sensory. They may be expressed in mental formulations—an idea or a statement— but they can never be enclosed within this formulation or in any way limited by our faculties. In the ancient traditional societies they were reflected, not merely in the theories whereby the mind organizes its material, but also in myths and rituals, as they were in every aspect of common life—man's waking and his sleeping, his eating, his love-making, his fighting and his work. This was the basis of that unity of life which most of us would give all that we have to repossess. Fragmentation of the personality is the salient characteristic of "modern" as against "primitive" man, and the problems which now arise regarding man's role in society, patterns of sexual behavior, or the distinction between creative work and servile labor are aspects of this fragmentation.

Since responsibility is necessarily a function of the whole man, those whose actions are regulated by only one part of their nature and who are at war with themselves find it easy to deny paternity when faced with the consequences of what they have done. The scientist whose pursuit of knowledge leads (indirectly, as it seems to him) to appalling consequences is aware that he never willed this outcome, very much as the man who rapes a young girl can say quite truly that he never meant to harm her. Scientists may suggest that the pursuit of knowledge for its own sake is natural to man, just as the rapist may feel that emotion, if it is powerful enough, contains its own justification; and both can take refuge in the excessive emphasis upon motives and conditioning which tends to isolate modern man from the great web of consequences which he actualizes. The fact remains that consequences do follow acts, and they must belong to someone.

There exists a popular image, fostered by the media, of the dedicated scientist, working long hours in his laboratory—yet happy as a child at play—careless about money and naïve in the ways of the world. Real scientists may not always be quite like this, but it is understandable if they adopt the required pose on occasion; like so many masks, it expresses a truth. When this "innocent" is faced with the consequences of his obsessive pursuit of knowledge, unregulated by any principle beyond a kind of mental lust, the truth becomes shockingly apparent. With indecent haste, he seeks for scapegoats (wicked politicians or rapacious businessmen) who have bent his precious discoveries to their own evil purposes. He had, of course, taken it for granted that none but angels would make use of the knowledge he has wrung from his intercourse with the natural world.

It is not as though he had never been warned; and this is the most astonishing aspect of the scientist's claim to innocence. The very fact that he is able to carry on his pursuit of factual knowledge is the outcome, at least according to one of the basic lessons children learn at school, of a long battle against "persecution," against "obscurantism" and against "superstition" or, in other words, against the massed weight of human opinion in earlier centuries. There is however another way of looking at the obstructions formerly placed in the way of scientific advance. A fence at the cliff's edge is an obstruction, certainly, but it has not been placed where it is without reason; and to suppose that the men who raised these obstructions were quite without intelligence or foresight is an impertinence which only reflects our own stupidity.

The battle against the physical sciences was waged with particular ferocity in Christendom at the end of the Middle Ages. The gestures of those who tried most desperately to halt the process make one think of dumb men attempting to prevent someone from striding cheerfully to perdition. The Inquisition, for example, did not have the right words, they could not have been expected to know the unknown or to see in detail where this new learning would lead, but a sound intuition alerted pious men to a fearful danger. In a fury of despair they would dig up a dead man's bones to condemn him, too late.

The investigation of the natural world "in depth" and the pursuit of factual knowledge for its own sake were then regarded as dangerous and ultimately destructive activities. It is absurd to be surprised if these activities do turn out to be both dangerous and ultimately destructive.

In the Islamic sector of the world the sciences showed less inclination to go off at a tangent from the total truth and were not subjected to the same "persecution." The presiding idea which dominates every aspect of Islamic thought—the divine Unity, beside which nothing can be said to have more than a shadowy and contingent existence—was of such power that fragmentary ideas were unlikely to escape from its magnetic field.

Even so, the note of warning was sounded often enough and Ibn 'Arabi, perhaps the greatest of the medieval Muslim philosophers, compared scientific delving into the secrets of nature to incest, a prying under the Mother's skirts; and this is one way of characterizing the desire of one facet of the natural world to know another in its most intimate contours. The penetration of nature by the fact-finding and analytic mind keeps time now with the rape of the earth we tread and with the exploitation of our fellow creatures. An incestuous conjunction of mind with matter engenders some monstrous offspring.

Our bodies (and there is a sense in which the whole world, the whole of nature, is our body) are clothing which lasts a little while and then falls apart. We have better things to do than pick obsessively at this clothing, placing its fragments under the microscope, making it our sole and absolute concern. Human dignity forbids such dreary obscenities.

It is not easy to stand out against the spirit of the age, nor is there any reason why it should be. It is right that people's minds should to some extent be closed to ideas which do not fit the framework of preconceptions which enables them to think and to act coherently; a man whose mind was wide open to every notion that came his way would be paralyzed by uncertainty and deafened by a cacophony of conflicting sounds. The fact remains that those who attempt to break down this protective wall of preconception start at an immense disadvantage when required to argue their case.

In the open societies of the West, free discussion and argument have great influence, particularly now that they are brought into almost every home by television. Where there exists a solid substratum of agreement—that is to say, whenever the debate is within the limits of the present climate of opinion—argument serves a practical purpose. If two men wish to travel to the same destination it is useful for them to argue over which is the best route to take; but if their goals are quite different they are bound to be at cross-purposes. Where there is a radical disagreement over fundamentals, argument, in the commonly accepted sense of the term, brings confusion rather than clarity. What do the opponent of science and the scientist—or, to come to essentials, believer and unbeliever—have to say to each other?

Not that dialogue is impossible. One can envisage a debate, held in quietness and intimate privacy (with no possibility of playing to the gallery), in which a believer and an unbeliever explore one another's minds over a long period and, inspired by a common desire to understand, achieve communication. Confined to thirty minutes in a television studio, such debate can only be farcical. Time and patience are of the essence, not to mention divine grace, love and a kind of stillness deeply infused with the longing for truth. Those who stand poles apart should never attempt hasty dialogue, unless they confine themselves to discussing the state of the weather. "Haste is from the devil," say the Muslims, "and slowness is from God"; and the clocks must be stopped if these two men are to understand each other.

But time is too valuable (when awareness of the timeless has been lost) for clocks to be stopped, love is at best a bit-player in this drama, and stillness is incompatible with controversy. All that such hasty debates between believer and unbeliever offer us is a battle of wits and a contest in verbal skills; and, since the former is out of tune with the spirit of his age, the rules of the game and the weapons are never of his choosing.

But perhaps there is no battle to be fought or won, for in most cases these antagonists have only the illusion of meeting and there is simply the spectacle—familiar in farce—of two men shadow-boxing on opposite sides of the stage, unaware that their blows never make contact. They are in different places. It is not enough to share a common language if there are no common assumptions to provide the basis for argument. Without any such basis each participant feels that the other is "missing the point," as indeed he is since the "point" is the truth as seen from the place at which he has taken his stand and these men are too far apart to share the same view. Heirs of a fairly unified culture, we still take a certain uniformity of viewpoint for granted, but in the modern age it is quite possible for people living side-by-side in the same society to inhabit entirely different worlds.

Because such a situation is by nature painful, those who take their stand upon the religious interpretation of the universe, being a minority and respecting democratic procedures, go to extraordinary lengths to meet their scientifically-minded companions rather more than half-way, as though a man tall enough to look over the fence were to squat down—for the sake of keeping company with his children—and peer through the hole they have bored in the wood, pretending this is all that can be seen of the next-door garden.

If "provincialism" is taken to indicate narrowness of viewpoint, then Eliade's phrase is particularly apt in relation to the contraction which has taken place over a long period and which was already well advanced among "educated" people when Descartes first made awareness of his own thinking self the starting-point of all human knowledge, taking care to shut the doors and windows before sinking into the cavern of mental self-awareness.

In appearance, the outer world has expanded as the inner has contracted. A small vaulted universe, lit by friendly lamps and haunted by familiar spirits, has opened out into the unimaginable vastness of space with its thin population of burning stars; on the other hand, inner space, a spiritual universe extending from nadir to Empyrean, has contracted to the dimensions of the skull-box. This process (reminiscent of the scientific theory of an "expanding universe") might be visualized in terms of a child's bubble-blowing—an "objective" world which swells in proportion to the life-breath pumped into it. But size, unless it has human significance, is nothing in relation to infinity. A distance of a thousand light-years is further than a man could walk; and having said this there is little more to be said about such distances. They are irrelevant to the business of being a man.

It is in this sense that man is "the measure of all things." As Viceroy, his concern is with the area given him as his particular and unique destiny; his only concern beyond this area is with an eternity subject neither to contraction nor expansion.

The vastness of "inner space," with its many dimensions, permits contraries to co-exist; but the attempt to fit truths which belong to different levels and make sense in terms of different perspectives into one framework at one particular level (that of the laws which govern our mental processes in the context of everyday life) is an impossible task. It is also an unnecessary task, for we ourselves do not live on one level only. But this is what rationalism, with its two-dimensional scheme of things, tries to do, and this is why the scientific view cannot be questioned on its own ground or in terms of the proofs and arguments which it considers valid.

It might seem too easy—and yet it would be true enough—to say that rationalism is false simply because it is an "-ism." It is false because of its pretensions to universality, its claim to include the whole of reality within its orbit, and because it excludes everything that cannot be fitted into its own particular and local categories; false, in other words, because it is a counterfeit, pretending to be something that it is not. Reason is one mode of knowledge among others, and rationalism is its "Pharaonic sin" (whereby the partial and fragmentary usurps the place of wholeness).

Man is a rational being, but he is also something much more. Reason is one of his tools—not his definition. Its nature is to operate in terms of irreconcilable alternatives: this is black or white; this creature is either male or female; either this animal will eat me or I shall eat it, and so on. These alternatives are indeed real on the level of our sensory experience, and since this experience is a form of true knowledge the instruments by means of which it is perceived and organized cannot be entirely false—so long as they keep their place. But the man who imagines that he can interpret all that *is* in terms of rational categories might be compared to someone who supposes he can absorb and digest knowledge through his belly.

Those who are unable to understand that they add up to more than the sum of their own instruments and who cannot accept the fact that the area of possible knowledge extends into moulds quite unrelated to the contours of the human mentality are prisoners in their own empirical and conditioned selfhood. Their speculation is a ball bounced against the walls of their cell.

The fact that there are aspects of truth which can never be formulated in logical terms is frustrating to the mind's lust for totality; the inconceivable is dismissed as being unknowable and therefore "unreal." Illusions are always conceivable, since they are rooted in our faculties and cannot exist without us; but truth does not need us and is independent of our faculties as it is of our powers of conceptualization. God, in his essence, is said to be quite inconceivable in terms of the mind's language; but there is nothing inconceivable about a flying hippopotamus, however improbable such a creature may be. The mind comprehends facts and is at ease with fictions. It is not by its nature apt to grasp realities unless enlightened by an enabling power which comes from beyond its sphere.

To dismiss partial modes of knowledge simply because they are what they are is just as grave a fallacy as to mistake the partial for something total and all-embracing. If reality could not in some measure be represented in mental, emotional and physical terms it would not be reality; and if the mind had no contact with reality we would all be mad. What has been lost in a mind-fixated age is awareness that the mental representation is by nature limited and incomplete, as is the emotional state or the physical image. Truth is expressed in these different languages without being exhausted by anything we can think, feel or say about it.

There is a necessary tension in the religious and intellectual spheres between acceptance and rejection of the partial images through which mind, emotion and senses maintain their hold on reality. Most of us cannot do without our mental concepts, our anthropomorphic image of God and our physical symbols; and the hidden truth responds to our need because it has its origin in the fountain of the divine Mercy and also because it is by nature partially conceivable, a fit object for love, and present in the sights, sounds, odors, flavors and tactile qualities of the physical world. To reject such partial knowledge as is offered by our natural faculties is a kind of self-mutilation; but to suppose that truth in its totality can be encompassed by these faculties is idolatry.

The inveterate human tendency to idolatry (worship of the reflection to the exclusion of that which is reflected) is, in the Muslim view, the most dangerous and the most universal of sins. The Islamic Revelation broke in upon a culture which had petrified into gross forms of idolatry; this was a moment in time when the breaking of images and the release of the spirit of truth from a stony prison was most necessary. But quite outside the historical circumstances which, providentially, determine the accent and emphasis of a particular religion, this Revelation had the function of redressing the balance between those who would bind the truth in mental formulae, emotional fixations and physical images, and those who insist upon its transcendence above all that we are capable of thinking or feeling or doing.

Without supernatural wisdom and without the humility which recognizes the subordination of the reasoning process to that wisdom, it is impossible for human minds to keep a just balance between transcendence and immanence, reconciling the notion of God as totally "other" (in Koranic terms, "having no likeness whatsoever") with the idea of God as intimately present everywhere ("closer to man than his jugular vein"); but it is still a useful exercise to set such contrary ideas side-by-side within the mind's narrow cabin (as do the Zen Buddhists by means of their paradoxical *koans*), until we begin to sense, far beyond our human reach, the existence of a point at which the contraries meet.

When two concepts, each capsulated in accordance with our mental needs, appear at once irreconcilable—as, for example, do the notions of predestination and free will—and yet necessary if our existence is to make any kind of sense, then we can only reach out towards that incomprehensible point. It is beyond the range of our bread-and-butter faculties, but this does not in any sense indicate that it is absent from the world or unrelated to the human person in his totality. On the contrary, the belief—normal to mankind—that there is a meaning inherent in everything that exists and everything that happens must necessarily imply the omnipresence of that point, that truth, that center.

Such beliefs as this are commonly classified as "mystical." They can then be treated, not with the hostility and resentment which so often accompanies the dismissal of "organized religion," but with a gesture of respect to a gentle and poetic eccentricity, too remote from everyday life to represent a threat to our way of life. And yet there have been some good swordsmen among the mystics who, like David, have slain their ten-thousands.

Insofar as the term has any precise meaning, mystics have no doubt followed their inward path in all places and all periods, triumphing over the obstacles presented by social chaos or social regimentation, sharing the vocation of the heroes and martyrs who stride over the turbulence or the petrifaction of their world with all the splendor of elephants rampaging through the bush. But the place they are going is the place we must all reach, and most people are not mystics, heroes or potential martyrs. They are not even elephants.

This is where the attempt to isolate mystical experience from the habitual stream of life in the sense in which, for example, musical experience may be isolated as something irrelevant to the lives of those who cannot share it, breaks down. The mystic is different from the rest only as the flyer is different from the walker, though both must hope to come to the city gate before nightfall. What he is talking about is also their business; but whereas he may find his way unaided by the society in which he happens to live, the common man, the quite unelephantine man, needs all the help he can get and has a right to expect this help from his society; and human societies, if they are to have any claim upon our loyalty beyond that of practical convenience, exist to beat a path through the bush for those who cannot fly or even trample.

What the traditional, God-centered societies offered their members was a life saturated with the awareness of realities beyond the reach of mind, feeling or sense in terms of their normal functioning and a whole system of bridges leading to mountain or hillock, as the case may be, but certainly leading outwards and upwards from the flatlands. The objects of sense were vivified by symbolism, emotion was universalized in ritual, and mental concepts were not self-sufficient propositions (limiting reality) but keys to supernatural knowledge.

In earlier times, says Thibon, "men did not know the contours of the human and cosmic lock, but they possessed the key. . . . Modern thought as a whole no longer occupies itself at all with the nature or existence of this key. The only question posed before a closed door is to examine it most painstakingly, not to open it."[3] Or else we ignore the door altogether (mistaking it for a section of an impenetrable wall) and set the key under a microscope, treating the instrument which lies in our hands as though it were an end in itself.

This could be a definition of idolatry: to worship a key instead of setting it to the lock. And here we come to the great divide which separates rationalism and all its offshoots from the traditional view of ideas, feelings and the phenomena of the material world as symbols and therefore as signs which, if properly used, point towards the timeless perfection which, in their flickering fashion, they signify. "We shall show them Our signs on the horizon and within themselves, until it is clear to them that *this* is the Truth." [Koran, 41:53] For the Islamic Revelation embodied in the Koran, all that we see and all that we find is of a superabundant richness, not on its own account, but because in its very existence it reflects the divine Qualities and reminds us of their source. A star, a bird on the wing, a forest or a river, and many lesser things ("Allah disdains not to coin the similitude even of a gnat . . .") are facets of a universal Revelation.

But to live with things that are other than they seem, among signs that point away from themselves, amidst bridges that lead elsewhere and ladders of which only the lower rungs are visible is hard for those who hunger after narrow certainties. It is easier to settle down where we are and regard the sign as a work of art, the bridge as a piece of masonry and the ladder as a wooden frame, accepting appearances for what little they are worth and trying to forget that death will—so far as we are concerned—reduce all such works to nothingness.

"Primordial man sees the 'more' in the 'less'," says Frithjof Schuon. "The infrahuman world in fact reflects the heavens and transmits in an existential language a divine message that is at once multiple and unique."[4] Christianity, he points out, could not fail to react against the real "paganism" of the cultural environment within which it crystallized as a world religion, but in so doing it also destroyed values which did not in the least deserve the reproach of

3 *Échelle de Jacob.* Gustave Thibon, p. 177.
4 *Images de l'Esprit.* Frithjof Schuon, pp. 15-16.

"paganism"; modern technology, he adds, "is but an end product, no doubt very indirect, of a perspective which, after having banished the gods and genies from nature and, having rendered it 'profane,' by this very fact finally made possible its 'profanation' in the most brutal sense of this word."

Paganism in the proper sense of the term is an idolatry applied to the natural world, but it is also, in most cases, the debris of a religion in the final stages of decay, when its adherents, like dogs, sniff at the pointed finger rather than going where the finger points; idolatry, animism, fetishism and other such aberrations (assuming that they exist objectively, and not merely in the modern observer's mind) all bear witness to the fact that phenomena which were once adored as symbols of transcendent realities have come to be worshipped for their own sakes. There are many intermediate stages in this degenerative process, and it follows that one cannot always mark the dividing line between images which are adored for what they symbolize and those which are worshipped as "gods." In any religious context—and most of all in that of Hinduism—there will be some men who understand that the image points away from itself and others who mistake it for an independent reality (in which case it becomes a counterfeit). At a time when the sacred is all but banished from our world, we do well to be tolerant of "superstition," so long as the intention behind it—a willingness to adore the holy—is sound.

A new divine Revelation, breaking in upon the rusty structure of the particular "milieu" into which it is directed, is likely to sweep the ancient images aside. It has no need of them. It offers a real and effective alternative, a highroad in place of the little paths and bridges which people had been using (or misusing) for ages past. But when the highroad itself has begun to suffer the erosion of time and has narrowed, then the loss is felt. Once it is out of sight—so far as the majority of people are concerned—no true path is to be despised, no bridge scorned as "naïve" or "childish."

It is, in any case, one thing for the lightning stroke to destroy such supports and quite another for busy, opinionated little men to set themselves up as wreckers.

Islam and Christianity were both, at their inception, revolutionary religions and therefore destructive, at least in a relative sense. It happens to have been that sector of the world which was formerly Christian that has imposed its patterns of thought and behavior almost universally, and ex-Christians are therefore the wreckers with whom we are chiefly concerned. The vast majority of Westerners who are not Christian believers in the full sense can fairly be described as ex-Christians (or pseudo-Christians). Such a heritage cannot easily be shaken off, and the fiercest opponents of religion are often those who cannot forgive God for not being a Christian (as they understand the term). The destructiveness which was once no more than a side effect of a great act of renewal turns sour and vicious in men for whom the blazing certainty of God's love and of Christ's redemptive sacrifice no longer have any meaning. The rose in decay stinks.

It could be said that the world is nothing but a tissue of bridges leading from here, where we find ourselves, to the "other shore," and in theory it is open to anyone to recognize sticks and stones for what they really are and so to discover a Paradise which was never finally lost. For him, no doubt, this world—so opaque, so darkened in this winter season—is still transparent as it is said to have been when it issued from the hand of God, and prison bars are no more than candy-sticks that snap in a child's grip. Perhaps there will always be such strangers, born out of their time, since time is not absolute and must sometimes be mocked. But what of the rest? The things we handle seem dark and heavy, the bars are thick, and age wears us out. We have great need of crutches and cannot be too proud to accept them wherever they are to be found. With them, we may hope to hobble over such rickety bridges as remain undestroyed.

What does a cripple feel, with fire or flood behind him and a jostling crowd making for the only exit, if someone wantonly knocks his crutch away and then destroys the bridge which led to safety? Rage, surely; and if men knew what they have lost through the arrogant destructiveness of the crutch-snatchers and bridge-wreckers their rage would make the anger of warring armies and revolutionary mobs seem kittenish.

The principal function of modern thought has been the wanton destruction of "superstition," a term which—though it may properly be applied to little habits which have survived in isolation from the doctrines which gave them meaning—has expanded to include every kind of belief in the supernatural. Bridges, ladders and, ultimately, the highroads provided by the great religions have at least one thing in common: they are invisible to those in whom this belief has been undermined.

It is difficult to measure wickedness and to define its degrees, but those who set themselves to persuade their fellow men that the world is nothing but a meaningless agglomeration of material particles (or the blind interaction of minute quanta of energy) totally separate from man's inner being and that there is no joy anywhere, no spiritual effort to be made, no eternal goal to be reached, have done a thing beside which no massacre of the innocents can stand comparison. Like the former Commandant of Auschwitz, these destroyers of bridges have been, for the most part, well-behaved, keeping their fingers off their neighbor's goods and their neighbor's wife; and this, as much as anything, makes current notions of morality seem infantile. If those who do the most harm go unpunished, how can we condemn mere thieves and murderers?

But if wickedness may often be defined in terms of a half-witted pursuit of relative good, then it can be said that much of this wrecking has been undertaken in the name of a fine ideal, the ideal of perfection. The idealist, the perfectionist, cannot tolerate what is grimy or flawed or broken. Lacking any sense of the sacred, lacking any courtesy towards creation and quite without modesty—a true "savage"—he rages to destroy the imperfect wherever he finds it (and that is everywhere). Our world is, by definition, a grimy, flawed and broken place, subject to decay and riddled with death. If it were otherwise it would not be the world or—to put the matter another way—this universe of time and space would be indistinguishable from the timeless and central perfection of Paradise and would therefore lose its separate existence. The world may be rendered transparent, so that perfection is discerned behind its shapes and patterns, and it may be loved so that its very deformities become the objects of a redeeming compassion; but it cannot be fundamentally changed at its own level. We have the power only to substitute one evil for another, "swapping black dog for monkey" (as the Jamaicans say) or leaping merrily out of the frying pan into the fire.

At the root of modern idealism, with its refusal to accept imperfection as something inherent in the human condition, there lies a bitter and perhaps satanic puritanism which, carried to its logical conclusion, would set fire to this world of ours and destroy it utterly.

"You can work miracles," said one of his companions to the Muslim saint, al-Hallaj; "Can you bring me an apple from Heaven?"

The saint raised his hand and, within the instant, held in it an apple which he offered to his friend. Biting into the fruit, the man observed with horror that there was a worm in it. "That," said al-Hallaj, "is because, in passing from the eternal realm into the world of time, it has taken on something of the latter's corruptibility."

This story has a bearing on contemporary attitudes to such traditional bridges as remain relatively intact in the modern world. Ignored or dismissed by the scientific view of reality, they are at the same time condemned for the rust which has settled on their outworks. They suffer the combined assault of rationalist and moralist. Even the man who has sufficient humility to acknowledge his own imperfections looks for a kind of primordial perfection in religious institutions and primordial purity in religious people. As a fallen being himself, he might be expected to know better.

Whatever "passes from the eternal realm into the world of time" must take on some of the limitations inherent in this world and become subject to the laws which govern the context of its incarnation. The organization and institutions in which a divine Revelation is fleshed cannot be immune to the process of decay, even though the grace which shines at the center of its manifestation remains untainted. Since we are what we are and the world is what it is, this reservoir of grace is tapped only by those who are prepared to embrace the outer shell until, like the Prince who awakens the Sleeping Beauty, they find what was always there, awaiting them in the innermost room of the castle. From this point of view the imperfections of any organized religion as it appears to the outsider and the scandal created by some of its representatives might be compared to the trials which the mythological Hero must surmount before he reaches the goal of all desire.

There are many people in our time who, with an arrogance which masks the inadequacies of a superficial education, think it intolerable that plaster saints and household icons and desert tombs should serve as bridges to true knowledge and that a God who is said to be almighty should permit his grace and power to operate through such seemingly humble instruments. They forget that this same God is omnipresent and that men are therefore apt to find him where they can.

The divine Presence within things—in sticks and stones, in bits and pieces—implies their wholeness, but those who are themselves inwardly divided and fragmented cannot recognize this. In the idealist's alienation, his refusal to stoop to the small, imperfect things, there is a profound betrayal of man's viceregality; for the Viceroy is a bridge-builder, and these men know only how to destroy. Obsessed by their ideas of neatness, they take their scissors and snip away at existence like a child who, when he tries to make his cut-out figure perfectly symmetrical, cuts first on one side, then on the other, until there is nothing left. They seek mastery through a process of reduction, and all that does not fit is to be eliminated; but in the long run nothing fits their categories. Everything must go.

"The explanation of the world by a series of reductions has an aim in view: to rid the world of extra-mundane values. It is a systematic banalization of the world undertaken for the purpose of conquering and mastering it. But the conquest of the world is not—in any case was not until half a century ago—the purpose of any human societies. It is an idiosyncrasy of Western man."[5]

In the traditional view of human destiny, degeneration is an inevitable feature of time and history; but this process can take quite different forms, on the one hand active and aggressive, tending to violence and, on the other, passive, indolent and relatively peaceful. One cannot doubt that the first of these is the white man's sickness and we know how contagious it has proved to be; but the possibilities inherent in human nature do not differ fundamentally between one race and another, and it might be more accurate to say that the white man brought out in Asians and Africans qualities which were already present, only waiting to be awakened—witness the speed with which Western vices and ideologies have spread through the rest of the world and also the eagerness with which so many traditional peoples have exchanged their own craftsmanship for Western junk. This final destructive fever had to break out somewhere. Once it had come to the surface, no sector of our world was immune.

The grim ambition to subdue creation to our own narrow purposes—symptomatic of the search for a counterfeit Paradise—is now almost universal. Its inevitable frustration must surely lead to increasing violence, ultimately self-destructive. And yet all this is no more than the frenetic activity of ants around their little mound, busy and blind under an indifferent sun. When they are done and

5 *The Two and the One*: Mircea Eliade (Harwell Press), pp. 156-157.

peace is over all, the sun will still be shining and the scattered fragments of existence will be re-assembled into the wholeness which is their only meaning: "And say—Truth has come and illusion has vanished away; illusion is indeed by nature ephemeral." (Koran, 17:81)

11

Ignorance

Wendell Berry

The expressed dissatisfaction of some scientists with the dangerous oversimplifications of commercialized science has encouraged me to hope that this dissatisfaction will run its full course. These scientists, I hope, will not stop with some attempt at a merely theoretical or technical "correction," but will press on toward a new, or a renewed, propriety in the study and the use of the living world.

No such change is foreseeable in the terms of the presently dominant mechanical explanations of things. Such a change is imaginable only if we are willing to risk an unfashionable recourse to our cultural tradition. Human hope may always have resided in our ability, in time of need, to return to our cultural landmarks and reorient ourselves.

One of the principal landmarks of the course of my own life is Shakespeare's tragedy of *King Lear*. Over the last forty-five years I have returned to *King Lear* many times. Among the effects of that play—on me, and I think on anybody who reads it closely—is the recognition that in all our attempts to renew or correct ourselves, to shake off despair and have hope, our starting place is always and only our experience. We can begin (and we must always be beginning) only where our history has so far brought us, with what we have done.

Lately my thoughts about the inevitably commercial genetic manipulations already in effect or contemplated have sent me back to *King Lear* again. The whole play is about kindness, both in the usual sense, and in the sense of truth-to-kind, naturalness, or knowing the limits of our specifically *human* nature. But this issue is dealt with most explicitly in an episode of the subplot, in which the Earl of Gloucester is recalled from despair so that he may die in his full humanity.

The old earl has been blinded in retribution for his loyalty to the king, and in this fate he sees a kind of justice for, as he says, "I stumbled when I saw" (*King Lear*, The Pelican Shakespeare, IV, i, 19). He, like Lear, is guilty of hubris or presumption, of treating life as knowable, predictable, and within his control. He has falsely accused and driven away his loyal son, Edgar. Exiled and under sentence of death, Edgar has disguised himself as a madman and beggar. He becomes, in that role, the guide of his blinded father, who asks to be led to Dover where he intends to kill himself by leaping off a cliff. Edgar's task is to save his father from despair, and he succeeds, for Gloucester dies at last " 'Twixt two extremes of passion, joy and grief." (V, iii, 199). He dies, that is, within the proper bounds of the human estate. Edgar does not want his father to give up on life. To give up on life is to pass beyond the possibility of change or redemption. And so he does not lead his father to the cliff's verge, but only *tells* him he has done so. Gloucester renounces the world, blesses Edgar, his supposedly absent son, and, according to the stage direction, "Falls forward and swoons" (IV, vi, 41).

When he returns to consciousness, Edgar now speaks to him in the guise of a passer-by at the bottom of the cliff, from which he pretends to have seen Gloucester fall. Here he assumes explicitly the role of spiritual guide to his father.

Gloucester, dismayed to find himself still alive, attempts to refuse help: "Away, and let me die" (IV, vi, 48).

And then Edgar, after an interval of several lines in which he represents himself as a stranger, speaks the filial (and fatherly) line about which my thoughts have gathered:

> Thy life's a miracle. Speak yet again.
> (IV, vi, 55)

This is the line that calls Gloucester back—out of hubris, and the damage and despair that invariably follow—into the properly subordinated human life of grief and joy, where change and redemption are possible.

The power of that line read in the welter of innovation and speculation of the bioengineers will no doubt be obvious. One immediately recognizes that suicide is not the only way to give up on life. We know that creatures and kinds of creatures can be killed, deliberately or inadvertently. And most farmers know that any creature that is sold has in a sense been given up on; there is a big difference between selling this year's lamb crop, which is, as such, all that it can be, and selling the breeding flock or the farm, which hold the immanence of a limitless promise.

*

A little harder to compass is the danger that we can give up on life also by presuming to "understand" it—that is by reducing it to the *terms* of our understanding and by treating it as predictable or mechanical. The most radical influence of reductive science has been the virtually universal adoption of the idea that the world, its creatures, and all the parts of its creatures are machines—that is, that there is no difference between creature and artifice, birth and manufacture, thought and computation. Our language, wherever it is used, is now almost invariably conditioned by the assumption that fleshly bodies are machines full of mechanisms, fully compatible with the mechanisms of medicine, industry, and commerce; and that minds are computers fully compatible with electronic technology.

This may have begun as a metaphor, but in the language as it is used (and as it affects industrial practice) it has evolved from metaphor through equation to identification. And this usage institutionalizes the human wish, or the sin of wishing, that life might be, or might be made to be, predictable.

I have read of Werner Heisenberg's principle that "Whenever one treats living organisms as physiochemical systems they must necessarily behave as such." I am not competent to have an opinion about the truth of that. I do feel able to say that whenever one treats living organisms as machines they must necessarily be perceived to behave as such. And I can see that the proposition is reversible: Whenever one perceives living organisms as machines they must necessarily be treated as such. William Blake made the same point earlier in this age of reduction and affliction:

> What seems to Be, Is, To those to whom
> It seems to Be, & is productive of the most dreadful
> Consequences to those to whom it seems to Be. . . .[1]

---

1 William Blake, *Complete Writings* (Oxford, 1966), p. 663.

For quite a while it has been possible for a free and thoughtful person to see that to treat life as mechanical or predictable or understandable is to reduce it. Now, almost suddenly, it is becoming clear that to reduce life to the scope of our understanding (whatever "model" we use) is inevitably to enslave it, make property of it, and put it up for sale.

This is to give up on life, to carry it beyond change and redemption, and to increase the proximity of despair.

Cloning—to use the most obvious example—is not a way to improve sheep. On the contrary, it is a way to stall the sheep's lineage and make it unimprovable. No true breeder could consent to it, for true breeders have their farm and their market in mind, and always are trying to breed a better sheep. Cloning, besides being a new method of sheep-stealing, is only a pathetic attempt to make sheep predictable. But this is an affront to reality. As any shepherd would know, the scientist who thinks he has made sheep predictable has only made himself eligible to be outsmarted.

The same sort of limitation and depreciation is involved in the proposed cloning of fetuses for body parts, and in other extreme measures for prolonging individual lives. No individual life is an end in itself. One can live fully only by participating fully in the succession of the generations, in death as well as in life. Some would say (and I am one of them) that we can live fully only by making ourselves as answerable to the claims of eternity as to those of time.

The problem, as it appears to me, is that we are using the wrong language. The language we use to speak of the world and its creatures, including ourselves, has gained a certain analytical power (along with a lot of expertish pomp) but has lost much of its power to designate *what* is being analyzed or to convey any respect or care or affection or devotion toward it. As a result we have a lot of genuinely concerned people calling upon us to "save" a world which their language simultaneously reduces to an assemblage of perfectly featureless and dispirited "ecosystems," "organisms," "environments," "mechanisms," and the like. It is impossible to prefigure the salvation of the world in the same language by which the world has been dismembered and defaced.

By almost any standard, it seems to me, the reclassification of the world from creature to machine must involve at least a perilous reduction of moral complexity. So must the shift in our attitude toward the creation from reverence to understanding. So must the shift in our perceived relationship to nature from that of steward to that of absolute owner, manager, and engineer. So even must our permutation of "holy" to "holistic."

At this point I can only declare myself. I think that the poet and scholar Kathleen Raine was correct in reminding us that life, like holiness, can be known only by being experienced.[2] To experience it is not to "figure it out" or even to understand it, but to suffer it and rejoice in it as it is. In suffering it and rejoicing in it as it is, we know that we do not and cannot understand it completely. We know, moreover, that we do not wish to have it appropriated by somebody's claim to have understood it. Though we have life, it is beyond us. We do not know how we have it, or why. We do not know what is going to happen to it, or to us. It is not predictable; though we can destroy it, we cannot make it. It cannot, except by reduction and the grave risk of damage, be controlled. It is, as Blake said, holy. To think otherwise is to enslave life, and to make, not humanity, but a few humans its predictably inept masters.

We need a new Emancipation Proclamation, not for a specific race or species, but for life itself—and that, I believe, is precisely what Edgar urges upon his once presumptuous and now desperate father:

> Thy life's a miracle. Speak yet again.

Gloucester's attempted suicide is really an attempt to recover control over his life—a control he believes (mistakenly) that he once had and has lost:

> O you mighty gods!
> This world I do renounce, and in your sights
> Shake patiently my great affliction off.
> (IV, vi, 34-36)

The nature of his despair is delineated in his belief that he can control his life by killing himself, which is a paradox we will meet again three and a half centuries later at the extremity of industrial warfare when we believed that we could "save" by means of destruction.

---

2 Kathleen Raine, *The Inner Journey of the Poet* (Braziller, 1982), pp. 180-181.

Later, under the guidance of his son, Gloucester prays a prayer that is exactly opposite to his previous one—

> You ever-gentle gods, take my breath from me;
> Let not my worser spirit tempt me again
> To die before you please
>
> (IV, vi, 213-215)

—in which he renounces control over his life. He has given up his life as an understood possession, and has taken it back as miracle and mystery. And his reclamation as a human being is acknowledged in Edgar's response: "Well pray you, father" (IV, vi, 215).

It seems clear that humans cannot significantly reduce or mitigate the dangers inherent in their use of life by accumulating more information or better theories or by achieving greater predictability or more caution in their scientific and industrial work. To treat life as less than a miracle is to give up on it.

*

I am aware how brash this commentary will seem, coming from me, who have no competence or learning in science. The issue I am attempting to deal with, however, is not knowledge but ignorance. In ignorance I believe I may pronounce myself a fair expert.

One of our problems is that we humans cannot live without acting; we have to act. Moreover, we *have* to act on the basis of what we know, and what we know is incomplete. What we have come to know so far is demonstrably incomplete, since we keep on learning more, and there seems little reason to think that our knowledge will become significantly more complete. The mystery surrounding our life probably is not significantly reducible. And so the question of how to act in ignorance is paramount.

Our history enables us to suppose that it may be all right to act on the basis of incomplete knowledge *if* our culture has an effective way of telling us that our knowledge is incomplete, and also of telling us how to act in our state of ignorance. We may go so far as to say that it is all right to act on the basis of sure knowledge, since our studies and our experience have given us knowledge that seems to be pretty sure. But apparently it is dangerous to act on the assumption that sure knowledge is complete knowledge—or on the assumption that our knowledge will increase fast enough to efface the bad consequences of the arrogant use of incomplete knowledge. To trust "progress" or our putative "genius" to solve all the problems that we cause is worse than bad science; it is bad religion.

A second human problem is that evil exists and is an ever-present and lively possibility. We know that malevolence is always ready to appropriate the means that we have intended for good. For example, the technical means that have industrialized agriculture, making it (by very limited standards) more efficient and productive and easy, have also made it more toxic, more violent, and more vulnerable—have made it, in fact, far less dependable if not less predictable than it used to be.

One kind of evil certainly is the willingness to destroy what we cannot make—life, for instance—and we have greatly enlarged our means of doing that. And what are we to do? Must we let evil and our implication in it drive us to despair?

The present course of reductive science—as when we allow agriculture to be invaded by the technology of war and the economics of industrialism—*is* driving us to despair, as witness the incidence of suicide among farmers.

If we lack the cultural means to keep incomplete knowledge from becoming the basis of arrogant and dangerous behavior, then the intellectual disciplines themselves become dangerous. What is the point of the further study of nature if that leads to the further destruction of nature? To study the "purpose" of the organ within the organism or of the organism within the ecosystem is *still* reductive if we do so with the assumption that we will or can finally figure it out. This simply captures the world as the subject of present or future "understanding," which will become the basis of further industrial and commercial optimism, which will become the basis of further exploitation and destruction of communities, ecosystems, and local cultures.

I am not of course proposing an end to science and other intellectual disciplines, but rather a change of standards and goals. The standards of our behavior must be derived, not from the capability of technology, but from the nature of places and communities. We must shift the priority from production to local adaptation, from innovation to familiarity, from power to elegance, from costliness to thrift. We must learn to think about propriety in scale and design, as determined by human and ecological health. By such changes we might again make our work an answer to despair.

12

The Plague of Scientistic Belief

Wolfgang Smith

Nothing strikes the contemporary mind as more certain and authoritative than the findings of physics, astronomy, chemistry, and, of late, molecular biology. These are the "hard" sciences of the present age, which, by empirical means, of a scope and accuracy that stagger the imagination, have put us in touch with fundamental realities that could not even have been conceived in bygone days. Moreover, this group of sciences has been in a sense "visibly validated," for all to see, by the technological miracles which now surround us on all sides; how, then, can one doubt—much less deny—their findings? In truth, one cannot; quantum particles and fields, galaxies and quasars, molecules and the genetic code—all these are undeniable facts, which must henceforth be reckoned with.

We must remember, however, that facts and their interpretation are not the same thing. And since, subjectively, facts are invariably associated with an interpretation of some kind, it comes about that science as a rule presents us with two disparate factors: with positive findings, on the one hand, plus an underlying philosophy in terms of which the formulation and disclosure of these discoveries are framed. In its actuality science is never the kind of purely empirical enterprise it is generally reputed to be, which is to say that ontological as well as epistemological presuppositions do inevitably play an essential role. What is more, these various philosophical articles of belief are rarely if ever examined or subjected to critical scrutiny by the scientific community. They are the foundational ideas one absorbs, as if by osmosis, in the course of one's scientific education; they pertain, one might almost say, to the scientific unconscious. And when it happens that one or the other of these ingrained philosophical dogmas does emerge into the light of day as a subject of discourse, the typical response on the part of scientists is to point immediately, by way of validation, to the success of the scientific enterprise: "It works!" one is told in effect. And yet in reality no philosophical belief has ever been validated by an empirical finding; the fact is that verification as well as falsification through empirical means apply to scientific as opposed to philosophical propositions. The separation between these two domains, however, is rarely attempted by scientists; only in times of extreme crisis, when the foundations of a science seem to be crumbling, does one encounter serious thought concerning questions of this kind, and even then such inquiries are pursued only by an adventurous few; it takes an Einstein or a Heisenberg to descend, as it were, to the foundational level, where philosophical axioms begin to come into view. What the rank and file absorb from these founders, moreover, pertains mainly to the technical aspect of the enterprise: one accepts the equations of relativity or the formalism of matrix mechanics, while all but ignoring the philosophical side of the coin. It is safe to say that the men and women who engage in the day-to-day business of scientific research tend not to be overly interested in philosophical subtleties; and so they incline to retain the philosophical axioms to which they have become accustomed over the years, and which could only be recognized as such, and dislodged, through serious and concentrated inquiry. It thus comes about that in the minds of scientists today, good science and inferior philosophy coexist and are in fact inextricably intertwined; as John Haught of Georgetown University has recently pointed out, "Some of the most prominent scientists are literally unable to separate science from their materialist metaphysics."

This said, I can proceed to state my primary thesis: I contend that by virtue of the aforesaid confusion scientists have promulgated philosophic opinions of the most dubious kind as established scientific truths, and in the name of science have thrust upon an awed and credulous public a shallow world-view for which in reality there is not a shred of scientific support. Having gained the trust and admiration of society through the technological wonders which they have engineered, I maintain that scientists as a class have usurped their authority by predisposing the public against the high truths of religion. I am not suggesting, to be sure, that they have consciously deceived others, but rather contend that as a rule they have themselves been misled in matters pertaining to philosophy, metaphysics, and religion. Meanwhile the fact remains that these "blind guides" are exerting an inestimable influence upon education and public belief, with disastrous consequences to human welfare, both here and hereafter.

*

I will apply the term "scientistic belief" to designate philosophical opinions that masquerade as scientific truths. Let me give two examples. As my first I will take the tenet of universal mechanism, or what could equally well be termed the axiom of physical determinism. The idea is simple: The tenet affirms that the external universe consists of matter whose motion is determined by the interaction of its parts. Given the initial configuration or state of this matter, and having once ascertained the laws which determine the effect of these interactions upon the resultant motion, one is supposedly able in principle to calculate the future evolution of the universe, down to the minutest detail. The cosmos is thus conceived as a kind of gigantic clockwork, in which part interacts with part to determine the movement of the whole. One knows that this idea began to take shape in the sixteenth century and has played a decisive role in the evolution of modern science. By the time of the Enlightenment, in fact, it had come to be almost universally regarded as an established scientific truth. Thus Hermann von Helmholtz, for instance, one of the leading scientists of the nineteenth century, could say with serene assurance: "The final goal of all natural science is to reduce itself to mechanics (*sich in Mechanik aufzulösen*)." With the advent of quantum theory, however, the picture has changed; for it turns out that the new physics is not compatible with the mechanistic premise. Yet, despite the fact of quantum indeterminism, not a few eminent scientists continue to champion the mechanistic tenet. Albert Einstein himself, as one knows, so far from admitting that the discoveries of quantum physics have overthrown the classical postulate, argued precisely in the opposite direction: it is the principle of determinism, he said in effect, that invalidates quantum mechanics as a fundamental theory. This illustrates quite clearly the philosophical and indeed *a priori* character of the tenet in question, and the fact that propositions of this kind can neither be verified nor falsified by empirical findings. This fact, however, remains generally unrecognized, with the result that the postulate of universal mechanism has retained to this day its status as a major article of scientistic belief.

My second example pertains to a more fundamental stratum of philosophical thought, and is consequently still more far-reaching in its implications: "physical reductionism," let us call it (for reasons which will presently become clear). The thesis hinges upon an epistemological assumption, an idealist postulate, one could say, which affirms that the act of sense perception terminates, not in an external object as we commonly believe, but in a subjective representation of some kind. According to this view, the red apple which we perceive exists somehow in our mind or consciousness; it is a subjective image, a fantasy which mankind has all along mistaken for an external object. Thus thought René Descartes, to whom we owe the philosophical foundations of modern science. Descartes sought to correct what he took to be the mistaken notions of mankind concerning perceptible entities by distinguishing between the external object, which he termed *res extensa*, and its subjective representation existing in the mind or so-called *res cogitans*. What was previously conceived as a single object (and what in daily life is invariably regarded as such) has therefore become split in two; as Whitehead has put it: "Thus there would be two natures, one is the conjecture and the other is the dream."[1] It is to be noted that this Cartesian differentiation between the "conjecture" and the "dream" goes not only against the common intuitions of mankind, but is equally at odds with the great philosophical traditions, including especially the Thomistic, where the opposition becomes as it were diametrical. Now, it is this questionable Cartesian doctrine—which Whitehead refers to as "bifurcation"—that has served from the start as the fundamental plank of physics, or better said, of the scientistic world-view in terms of which we habitually interpret the results of physics. And once again we find that the two disparate factors—the operational facts of physics and their customary interpretation—have become in effect identified, which is to say that the tenet of bifurcation does indeed function as a scientistic belief.

---

1 *The Concept of Nature* (Cambridge University Press, 1964), p. 30. Despite his eminence as a philosopher and the fact that, along with Bertrand Russell, he is the father of mathematical logic, Whitehead's strictures against the Cartesian axioms have aroused little response from the scientific community.

I would like to emphasize that in addition to the fact that bifurcation contradicts the most basic human intuitions as well as the most venerable philosophical traditions, there is also not a shred of empirical evidence in support of this heterodox position. Nor can there be, as follows from the fact that physics can be perfectly well interpreted on a non-bifurcationist basis, as I have shown in a recent monograph.[2] It turns out, moreover, that the moment one does interpret physics in non-bifurcationist terms, the so-called quantum paradoxes—which have prompted physicists to invent the most bizarre ontologies—vanish of their own accord. It seems that quantum physics has thus implicitly sided with the pre-Cartesian world-view.

It remains to explain why I have referred to bifurcation as "physical reductionism." The reason becomes clear the moment we return to the bedrock of the perennial *Weltanschauung*. The red apple we perceive belongs then once more to the external world; it constitutes a corporeal object, I will say, meaning thereby that it can be perceived. The "molecular" apple, on the other hand, with which the physicist is concerned, is bereft of sensible qualities, and is consequently imperceptible. It constitutes what I term a physical object, as distinguished from a corporeal. From a bifurcationist point of view, however, the physical object is all that exists in the external world. The corporeal, thus, is conceived in effect to be "nothing but" the physical. The red apple—which, from an orthodox point of view, exists!—is thus in effect "reduced" to the physical: it is identified with the "molecular" apple, as conceived by the physicist. The tenet of bifurcation, therefore, implies what I term physical reductionism; and the converse, to be sure, is equally apparent.

In both of these two forms, the Cartesian thesis has been for centuries presupposed without question by scientists and the educated public. It has become ingrained in the scientific mind to the point where even the anomalies of quantum physics have failed to arouse suspicion. As one philosopher of science has recently

2 *The Quantum Enigma* (Peru, Illinois: Sherwood Sugden, 1995). A useful summary of the book with commentary has been given by William A. Wallace in "Thomism and the Quantum Enigma," *The Thomist* 61 (1997), pp. 455-467. See also Wolfgang Smith, "From Schrödinger's Cat to Thomistic Ontology," *The Thomist* 63 (1999), pp. 49-63.

admitted in private: "Those who work on the physicist's plane find it almost impossible to eliminate the bifurcationism implicit in their work." Now, this uncritical and habitual acceptance of the Cartesian thesis by "those who work on the physicist's plane" effectively obscures its philosophical status; and as is the case with all scientistic beliefs, the tenet thus becomes science by association, as one might say.

One could argue that bifurcation—or, equivalently, physical reductionism—constitutes in fact the most basic contemporary scientistic belief, the tenet which all other scientistic beliefs implicitly presuppose. Take, for instance, the idea of universal mechanism: does it not hinge upon bifurcation? In a remarkable passage, amply worth quoting, Descartes himself admits as much:

> We can easily conceive how the motion of one body can be caused by that of another, and diversified by the size, figure and situation of its parts, but we are wholly unable to conceive how these same things can produce something else of a nature entirely different from themselves, as for example, those substantial forms and real qualities which many philosophers suppose to be in bodies.[3]

The philosophers alluded to, of course, are the Scholastics, whom Descartes opposes radically. What the French savant tells us—with admirable clarity!—is that not until the universe has been reduced to the status of "quantified matter" does the idea of universal mechanism become conceivable. And is this not, finally, the reason why Galileo and Descartes saw fit to ban "those substantial forms and real qualities" from the external world? Was not the bifurcation postulate introduced precisely to render thinkable a "totalist" physics based upon mechanical principles?

The two examples may suffice to introduce the general phenomenon which I have termed scientistic belief. It hardly needs pointing out, moreover, that if physics, the most exact of the natural sciences, is thus associated with scientistic—and indeed, from a traditional point of view, illusory!—notions, what can one expect in the case of less rigorous disciplines, such as evolutionary biology, physical anthropology, and psychology, not to speak of the so-called

---

3 Cited in E. A. Burtt, *The Metaphysical Principles of Modern Physical Science* (New York: Humanities Press, 1951), p. 112.

social sciences.[4] The unappreciated fact is that science in its actuality bestows both truth and error: not only enlightenment, but benightedness as well. One could even argue that so far as the general public is concerned, it is the second of these effects that predominates; the truths of hard science, after all, are mainly accessible to the expert, the scientifically proficient. This holds especially in the case of fundamental physics; by the time a fact of quantum theory, for instance, has been popularized, what remains is mainly a scientistic notion. One could put it this way: As science evolves, its actual insights become more and more abstract, more and more mathematical, and thus denuded of sensible imagery; these insights thus become a kind of esoteric knowledge, to which only the "initiated" have access. Moreover, what is validated by empirical findings, and also, in a way, by the miracles of technology, is precisely that kernel of esoteric insight, and *not* the outer shell of scientistic beliefs, which the public at large mistakes for enlightenment.

*

I would like now to consider the implications of these facts—of this cultural phenomenon—with reference to religion and the spiritual life. As has already been noted, I perceive the impact of scientistic belief upon the religious domain as adverse in the extreme. I should add that the problem has been greatly exacerbated by the fact that theologians and pastors as a rule are ill-equipped to deal with questions of this kind, and all too often have themselves been swayed by scientistic claims.

What does it matter, some will say; what if we are perhaps mistaken about the nature of causality, or about the terminus of sense perception, or even about the much-debated question of evolution—so long as we stand on the side of truth in matters of religion. I would point out that the question is not quite so simple. We must not forget that religion—so long as it has not degenerated into a social convention or mere sentimentality—demands the whole man; holiness and wholeness are inseparable. Does not the "first and greatest" commandment enjoin that "Thou shalt love the Lord

---

4 See *Cosmos and Transcendence* (Peru, Illinois: Sherwood Sugden, 1984), a work in which I have sought to unmask the major articles of scientistic belief and delineate their impact upon contemporary society.

thy God with all thy heart, and with all thy soul, and *with all thy mind*"? What we think about the world—our *Weltanschauung*—cannot legitimately be excluded from the domain of religion. As St. Thomas Aquinas writes in the *Summa Contra Gentiles* (Bk. II, ch. 3): "It is absolutely false to maintain, with reference to the truths of our faith, that what we believe regarding the creation is of no consequence, so long as one has an exact conception concerning God; because an error regarding the nature of creation always gives rise to a false idea about God." I would add that I perceive the contemporary penchant for accommodating the teachings of Christianity to the so-called truths of science as a striking confirmation of this Thomistic principle: a case, almost invariably, of scientistic errors begetting flawed theological ideas.[5]

In a word, what we think about the universe does matter in our religious and spiritual life. And moreover, with due allowance for what might be termed "invincible ignorance," we are responsible for the opinions we hold in this seemingly secular domain. "With all thy mind": these four words should suffice to apprise us of this fact.

I will go so far as to contend that religion goes astray the moment it relinquishes its just rights in the so-called natural domain nowadays occupied by science. I believe that the contemporary crisis of faith and the ongoing de-Christianization of Western society have much to do with the fact that for centuries the material world has been left to the mercy of the scientists. This has of course been said many times before (yet not nearly often enough!). Theodore Roszak, for one, has put it exceptionally well: "Science is our religion," he observed, "because we cannot, most of us, with any living conviction see around it."[6] And one might add that perhaps only those who already have at least a touch of authentic religion do in fact stand a chance of "seeing around it with any living conviction." So too the name of Oskar Milosz (1877-1939) comes to mind, a European writer who had this to say: "Unless a man's concept of the physical universe accords with reality, his spiritual life will be crippled at its roots, with devastating

---

5 The paramount instance of scientistic theology is doubtless given by the far-flung speculations of Teilhard de Chardin. See my monograph, *Teilhardism and the New Religion* (Rockford, Illinois: TAN Books, 1988), where I have dealt with this question at length.

6 *Where the Wasteland Ends* (Garden City: Doubleday, 1973), p. 124.

consequences for every other aspect of his life."[7] It could not have been better said! As regards the implications of the scientistic world-view for the life of the Church, let me quote from a recent book by the French philosopher Jean Borella: "The truth is that the Catholic Church has been confronted by the most formidable problem a religion can encounter: the scientistic disappearance (*disparition scientifique*) of the universe of symbolic forms which enable it to express and manifest itself, that is to say, which permit it to exist." And he goes on to say: "That destruction has been effected by Galilean physics, not, as one generally claims, because it has deprived man of his central position—which, for St. Thomas Aquinas is cosmologically the least noble and the lowest—but because it reduces bodies, material substance, to the purely geometric, thus making it at one stroke scientifically impossible (or devoid of meaning) that the world can serve as a medium for the manifestation of God. The theophanic capacity of the world is denied."[8] Let us be clear about it: Borella is pointing the finger squarely at what I have termed physical reductionism: "*le problème le plus redoubtable qu'une religion puisse rencontrer,*" he calls it. What he terms a "reduction to the purely geometric" corresponds precisely to what I call the reduction of the corporeal to the physical: it is this scientistic contention that would obliterate "the theophanic capacity of the world."

It is of course to be understood that the "symbolic forms" to which Borella refers are not, as some might think, subjective images or ideas which in days gone by men had projected upon the external universe, until, that is, science came to apprise us of the truth. The very opposite is in fact the case: The "forms" in question are objectively real and indeed essential to the universe. We may conceive of them as "forms" in the Aristotelian and Scholastic sense, or Platonically, as eternal archetypes reflected on the plane of corporeal existence. In either case they constitute the very essence of corporeal being. Remove these "symbolic forms," and the universe

7 Cited in Seyyed Hossein Nasr, *Religion and the Order of Nature* (Oxford University Press, 1996), p. 153. Concerning Oskar Milosz, see Philip Sherrard, *Human Image: World Image* (Ipswich: Golgonooza Press, 1992), pp. 131-146.

8 *Le sens du surnaturel* (Geneva: Editions Ad Solem, 1996), p. 74. See also the English translation: *The Sense of the Supernatural* (Edinburgh: T & T Clark, 1998).

ceases to exist; for it is these "forms," precisely, that anchor the cosmos to God.

It is needless to point out that science has not in reality destroyed these forms, or caused their disappearance; however, the scientistic negation of corporeal being entails a denial of the substantial forms or essences which constitute that order of being, and of the sensible qualities by which these forms or essences manifest themselves to man. The scientistically prepared mind, therefore, has become increasingly insensitive to what Borella terms "the universe of symbolic forms," to the point where that universe has become for it all but invisible. It is in that sense that the "theophanic capacity of the world" has been diminished to an unprecedented degree.

The consequences, however, of that diminution cannot but be tragic in the extreme. In his denial of essences, scientistic man has destroyed the very basis of the spiritual life. As Borella points out, he has obliterated the domain "that enables the Church to express and manifest itself," and hence "permits it to exist." The refutation of scientistic belief, therefore, is not an optional matter for the Church, something from which she can afford to abstain; it is rather a matter of urgent necessity, a question ultimately of survival.

It may be well, finally, to reflect anew upon what St. Paul has to say concerning "the theophanic capacity of the world" in his letter to the Romans. "For the invisible things of him from the creation of the world are clearly seen," he declares, "being understood by the things that are made, even his eternal power and Godhead." To which he adds: "So they are without excuse: Because that, when they knew God, they glorified him not as God, neither were they thankful; but became vain in their imaginations, and their foolish heart was darkened. Professing themselves to be wise, they became fools." (Romans 1:20-22) I need hardly point out the striking relevance of these words to all that we have discussed. The "things that are made" are doubtless corporeal natures, the objects that man can perceive; and what about "the invisible things of him": are these not precisely eternal essences, ideas or archetypes? So long as man's heart has not been "darkened," the sensory perception of "things that are made" will awaken in him an intellectual perception—a "recollection," as Plato says—of the eternal things which the former reflect or embody. St. Paul alludes to a time or a state when man "knew God," a reference, first of all, to the condition of Adam before the fall, when human nature was as yet undefiled by original sin. One needs to realize, however, that the fall of Adam has been repeated on a lesser scale down through the ages, in an unending series of "betrayals," large and small. Even today, at this late stage of history, we are, each of us, endowed with a certain "knowledge of God" to which we can freely respond in various ways. And that is precisely why we, too, are "without excuse," and why, to some degree at least, we are responsible for the opinions we hold concerning the cosmos. Everyone perceives the universe in accordance with his spiritual state: the "pure in heart" perceive it without fail as a theophany; and for the rest of us, whose "foolish hearts are darkened," the theophanic capacity of the universe is reduced in proportion to this darkening.

I would like however to emphasize that this correspondence between our spiritual state and our *Weltanschauung* applies in both directions, which is to say that not only does our spiritual state affect the way we view the external world, but conversely, our views concerning the universe react invariably upon that state. This is in fact my central point: Cosmology matters, it has a decisive impact upon our spiritual condition. Even what we think about the purely physical world turns out to be crucial; for indeed, "unless a man's concept of the physical universe accords with reality, his spiritual life will be crippled at its roots. . . ."

*

This brings us at last to the pastoral question: what can be done pastorally to counteract the scientistic influence? The major problem, clearly, is to inform the pastors themselves: to alert them, first of all, to the fact that there is a crucial distinction to be made between science and scientism, and then to the fact that scientistic belief is antagonistic to our spiritual well-being. This however will not be easy to get across, for it offends against the prevailing trend, both in civil society and within the Church. It is only by an act of grace, I surmise, that any of us are able to muster the discernment, and indeed the sheer boldness, to cast off the scientistic *Weltanschauung* and recover a Christian world-view. And this task, this imperative, I say, is at bottom spiritual. It is to be accomplished, thus, not simply by reading books, or through a process of reasoning, but above all through faith and prayer. The dictum *credo ut intelligam* applies to us still, and perhaps even more urgently than in the comparatively innocent days of Augustine or Anselm. It is needful that we be touched and enlivened by the Holy Ghost, the Spirit of truth, who "will guide you into all truth." (John 16:13) In our struggle to transcend the scientistic outlook, we are dealing, moreover, not simply with a belief system of human contrivance, but with something more formidable by far; for here too, in the final count, "we wrestle not against flesh and blood, but against principalities, against powers, against the rulers of the darkness of the world, against spiritual wickedness in high places." (Eph. 6:12) How could it be otherwise when it is "the theophanic capacity of the world" that stands at issue: the very thing "which enables the Church to express and manifest itself, that is to say, which permits it to exist." If the cosmos were indeed what scientism affirms it to be, our Catholic faith would be a mockery, and our sacred liturgy—the well-spring of the Church itself—an empty charade. This fact cannot be ignored with impunity.

13

Scientism: The Bedrock of the Modern Worldview

Huston Smith

Only four letters, "tism," separate scientism from science, but that small slip twixt the cup and the lip is the cause of all our current problems relating to worldview and the human spirit. Science is on balance good, whereas nothing good can be said for scientism.

Everything depends on definitions here, for this chapter will fall apart if the distinction between *science* and *scientism* is allowed to slip from view. To get those definitions right requires cutting through the swarm of thoughts, images, sentiments, and vested interests that circle the word *science* today to arrive at the only definition of the word that I take to be incontrovertible—namely, that science is what has changed our world. Accompanied by technology (its spin-off), modern science is what divides modern from traditional societies and civilizations. Its content is the body of facts about the natural world that the scientific method has brought to light, the crux of that method being the controlled experiment with its capacity to winnow true from false hypotheses about the empirical world.

Scientism adds to science two corollaries: first, that the scientific method is, if not the *only* reliable method of getting at truth, then at least the *most* reliable method; and second, that the things science deals with—material entities—are the most fundamental things that exist. These two corollaries are seldom voiced, for once they are brought to attention it is not difficult to see that they are arbitrary. Unsupported by facts, they are at best philosophical assumptions and at worst merely opinions. This book[1] will be peppered with

---

1 *Editor's note:* This chapter is taken from the book: *Why Religion Matters: The Fate of the Human Spirit in an Age of Disbelief,* (San Francisco: Harper, 2000).

instances of scientism, and one of Freud's assertions can head the parade: "Our science is not illusion, but an illusion it would be to suppose that what science cannot give us we can get elsewhere." Our ethos teeters precariously on sandy foundations such as this.

So important and undernoticed is this fact that I shall devote another paragraph to stating it more concretely. For the knowledge class in our industrialized Western civilization, it has come to seem self-evident that the scientific account of the world gives us its full story and that the supposed transcendent realities of which religions speak are at best doubtful. If in any way our hopes, dreams, intuitions, glimpses of transcendence, intimations of immortality, and mystical experiences break step with this view of things, they are overshadowed by the scientific account. Yet history is a graveyard for outlooks that were once taken for granted. Today's common sense becomes tomorrow's laughingstock; time makes ancient truth uncouth. Einstein defined common sense as what we are taught by the age of six, or perhaps fourteen in the case of complex ideas. Wisdom begins with the recognition that our presuppositions are options that can be examined and replaced if found wanting.

The Flagship Book

My flagship book for this chapter is Bryan Appleyard's *Understanding the Present: Science and the Soul of Modern Man*. I will compress its thesis into a story, the details of which are mine, but whose plot is his.

Imagine a missionary to Africa. Conversion is slow going until a child comes down with an infectious disease. The tribal doctors are summoned, but to no avail; life is draining from the hapless infant. At that point the missionary remembers that at the last minute she slipped some penicillin into her travel bags. She administers it and the child recovers. With that single act, says Appleyard, it is all over for the tribal culture. Elijah (modern science) has met the prophets of Baal, and Elijah has triumphed.

If only that tribe could have reasoned as follows, Appleyard continues; if only they could have said to themselves: This foreigner obviously knows things about our bodies that we do not know, and we should be very grateful to her for coming all this distance to share her knowledge with us. But as her medicine appears to tell us nothing about who we are, where we came from, why we are here, what we should be doing while we are here (if anything), and what happens to us when we die, there seems to be no reason why we cannot accept her medicine gratefully while continuing to honor the great orienting myths that our ancestors have handed down to us and that give meaning and motivation to our lives.

If only those tribal leaders had the wit to reason in that fashion, Appleyard concludes, there would be no problem. But they do not have that wit, and neither do we.

From that fictionalized condensation of Appleyard's book, I proceed to develop its thesis in my own way, beginning with the reception his book received.

Before I had laid hands on Appleyard's book, I attended a conference at the University of Notre Dame. Finding myself at breakfast one morning with the noted British scientist Arthur Peacocke, I asked him about the book, for it had first appeared in England and I thought Peacocke might have gotten the jump on me in reading it. He said that he had not read it but had heard that it was an anti-science book.

Click! Scientism. Scientism, because when I got to the book it turned out not to be against science at all, not science distinct from scientism. But because it spells out with unusual force and clarity what social critics have been saying for some time now—namely that we have turned science into a sacred cow and are suffering the consequences idolatry invariably exacts—it is a sitting duck to be taken as an attack on the scientific enterprise. Not by all scientists. It is not a digression to say (before I continue with Appleyard) that not all scientists idolize their profession. The spring 1999 issue of the *American Scholar* that crosses my desk on the day that I write this page bears this out forcefully. Its review of *Of Flies, Mice, and Men* sees its author, the French microbiologist François Jacob, as having written his book "to renounce much of the epistemological privilege of science, for as [he] points out with surprising and even extreme determination, the myths, misconceptions, and misuses of science can be insidious. They infiltrate our language and beliefs even as we try to expel them."

I could hardly ask for a stronger ally in this chapter than biologist Jacob, and with his support I return to Bryan Appleyard.

When *Understanding the Present* was published, responses to it polarized immediately. The *Times Literary Review* saw the book's author as voicing truths that needed to be spoken, whereas

England's leading scientific journal, *Nature*, branded it "dangerous."

When reviews began to appear on this side of the Atlantic, the *New York Review of Books* chose a science writer, Timothy Ferris, to do the job. Ferris gives us his opinion of the book in his closing paragraph. "Its real target," he writes, "would appear to be not science but scientism, the belief that science provides not *a* path to truth, but the *only* path." So far, fair enough—but then Ferris tells us that:

> Scientism flourished briefly in the nineteenth century, when a few thinkers, impressed by such triumphs as Newtonian dynamics and the second law of thermodynamics, permitted themselves to imagine that science might soon be able to predict everything, and we ought to be able to muster the sophistication to recognize such claims as hyperbolic. Scientism today is advocated by only a tiny minority of scientists.

Those of us who stand outside the science camp can only read such words with astonishment. "Scientism flourished *briefly* when *a few* thinkers permitted themselves to imagine that science might soon be able to predict everything"? "Scientism today is advocated by only *a tiny minority* of scientists"? Ferris's assertions dismiss *the* metaphysical problem of our time by definitional fiat, for if you define scientism as the belief "that science might soon be able to predict everything," then of course too few people believe *that* for it to constitute a problem.

Tracking Scientism

A discussion I was party to recently comes to mind. Historians of religion were asking themselves why the passion for justice surfaces more strongly in the Hebrew scriptures than in others, and when someone came up with the answer it seemed obvious to us all. No other sacred text was assembled by a people who had suffered as much *injustice* as the Jews had, and this made them privy from the inside to the pain injustice occasions. It is extravagant to compare the damage that scientism wreaks to the suffering of the Jews, but the underlying principle is the same in both cases. Only discerning victims of scientism (and sensitive scientists like François Jacob whom I quoted several paragraphs back) can comprehend the magnitude of its oppressive force and the problems it creates. For it takes an eye like the one Michel Foucault trained on prisons, mental institutions, and hospitals (which eye I am striving for in this book) to detect the power plays that the micro-practices of scientism exert in contemporary life.

Another procedural point must be entered, for it too is often overlooked. What is and is not seen to be scientism is itself metaphysically controlled, for if one believes that the scientific worldview is true, the two appendages to it that turn it into scientism are not seen to be opinions. (I remind the reader that the appendages are, first, that science is our best window onto the world and second, that matter is the foundation of everything that exists.) They present themselves as facts. That they are not provable does not count against them, because they are taken to be self-evident—as plainly so as the proverbial hand before one's face.

This poses *the* major problem for this book, because what is taken to be self-evident depends on one's worldview, and disputes among worldviews are unresolvable. Today's science-backed self-evidence is a fact of contemporary life that must be lived with. It is like wind in one's face on a long journey: to be faced without allowing it to divert one from one's intended course. During the McCarthy era it was said that Joe McCarthy found Communists under every bed, and those who are on the science side in this debate will see me as doing the same with scientism—or as finding under stones the sermons I have already put there, as Oscar Wilde charged Wordsworth with doing. There being (from their point of view) no problem, they will see this entire book as an exercise in paranoia. Because the difference comes down to one of perception, I will plow ahead in the face of that charge, taking heart from the way Peter Drucker perceived his vocation.

As the dean of management consultants in their founding generation, Drucker received every honor that his field had to confer. When he retired, he was asked in an interview if there was anything professionally that he would have liked to have had happen that had not happened. Drucker answered that actually there was. He kept replaying in his mind a scenario that in real life had never transpired. In it he was seated with the CEO of a company in the wrap-up session of a two-week consultation. Having looked together into every aspect of the company's operations they could think of, the two had become friends and grown used to speaking frankly to each other, so at one point the CEO leans back in his chair and says, "Peter, you haven't told me a thing I didn't already know."

"Because," Drucker added, "that's invariably the case. I never tell my clients anything they don't already know. My job is to make them see that what they have been dismissing as incidental evidence is actually crucial evidence." That is what I see myself doing with respect to scientism in this book.

Having referred to the *New York Review of Books* regarding its handling of Appleyard's book, I will turn to it again for my next example of scientism, for that journal serves as something of a house organ for the elite reading public in America.

John Polkinghorne is a ranking British scientist who at the age of fifty became an Anglican clergyman. The *New York Review of Books* never reviews theological books; but presumably because Polkinghorne is also a distinguished scientist, it made an exception in his case. To review his book, the *NYRB* reached for a world-class scientist, Freeman Dyson. *Click!* A scientist to review a book on theology? To see what that choice bespeaks, we need only turn the table and try to imagine the editors of the *NYRB* reaching for a theologian to review a book on science. The standard justification for this asymmetry is that science is a technical subject whereas theology is not, but now hear this. Several years back at a conference at Notre Dame University I heard a leading Thomist say in an aside to the paper he was delivering, "There may be—there just *may* be—twelve scholars alive today who understand St. Thomas, and I am not one of them."

We turn now to what Dyson said about Polkinghorne's book. After commending its author for his contributions to science and for historical sections of the book under review, Dyson turned to his theology, which like all theology, he said, suffers from being about words only, whereas science is about things. *Click* and *double-click!* As a self-appointed watchdog on scientism, I took pen in hand and challenged that claim in a letter to the *NYRB* that began as follows:

> It is symptomatic of the unlevel playing field on which science and religion contend today that a scientist with no theological credentials (Freeman Dyson in the *New York Review*, May 28, 1998) feels comfortable in concluding that the theology of a fellow scientist (John Polkinghorne) is, like all theology, about words and not, as is the case with science, about things. This flies in the face of the fact that most theology takes God to be the only completely real "thing" there is, all else being like shadows in Plato's cave. Muslims in their testament of faith sometimes transpose "There is no God

but the God" to read, "There is no Reality but *the* Reality," the two assertions being identical.

The rest of my letter is irrelevant here, but I do want to quote the first sentences of Freeman Dyson's reply as indicative of the graciousness of the man. "I am grateful to Huston Smith for correcting my mistakes," he wrote. "I have, as he says, no theological credentials. I have learned a lot from his letter." Dyson may have no theological credentials, but he is certainly a gentleman.

In a chapter that has to struggle at every turn not to sound peevish and aggrieved, whimsy helps, so I will mention the occasion on which I found scientism aimed most pointedly (though disarmingly) at me. (I told the story in my *Forgotten Truth*, but it bears repeating here.)

Not surprisingly, the incident took place at MIT, where I taught for fifteen years. I was lunching at the faculty club and found myself seated next to a scientist. As often happened in such circumstances, the conversation turned to the differences between science and the humanities. We were getting nowhere when suddenly my conversational partner interrupted what I was saying with the authority of a man who had discovered Truth. "I have it!" he exclaimed. "The difference between us is that I count and you don't." Touché! Numbers being the language of science, he had compressed the difference between C. P. Snow's "two cultures" into a double entendre.

The tone in which his discovery was delivered—playful, but with a point—helped, as it did on another MIT occasion. When I asked a scientist how he and his colleagues regarded us humanists, he answered affably, "We don't even bother to ignore you guys." Despite the levity in these accounts, the very telling of them opens me to the charge of sour grapes, so to those who will say that I am embittered I will say that they are quite wrong. Our scientific age has, if anything, treated me personally above my due. My concern is with scientism's effect on our time, our collective mindset—the fact that (to go back to Appleyard) it is "spiritually corrosive, and, having wrestled religion off the mat, burns away ancient authorities and traditions." The chief way it does this, Appleyard continues, "is by separating our values from our knowledge of the world." Timothy Ferris dismisses this charge as "extravagant and empty," and here again we can only be astonished at how blind those inside the scientific worldview are to the scientism that others find riddling modernism throughout. For, science writer that he is, there is no way Ferris could have been unaware that Jacques Monod drew a gloomier conclusion from our having separated values from knowledge than Appleyard does. Think of one of the key assertions by Monod: "No society before ours was ever rent by contradictions so agonizing. . . . What we see before us is an abyss of darkness."

Thus far this chapter has proceeded largely in the wake of Appleyard's book. I want soon to strike out on my own, but not before adding Appleyard's most emphatic charge, which is that "science has shown itself unable to coexist with anything." Science swallows the world, or at least more than its share of it. Appleyard does not mention Spinoza in this connection, but I find in Spinoza's *conatus* the reason for Appleyard's charge.

Spinoza's *Conatus*

Spinoza wrote in Latin, and the Latin word *conatus* translates into English as "will." Every organism, Spinoza argued, has within it a will to expand its turf until it bumps into something that stops it, saying to it, in effect, *Stay out; that's my turf you're trespassing on.* Spinoza did not extend his point to institutions, but it applies equally to them, and I find in this the explanation for why science has not yet learned the art of coexistence. Most scientists as individuals have mastered that art, but when they gather in institutions—the appropriately named American Association for the Advancement of Science, the *Scientific American*, and the like—collegiality takes over and one feels like a traitor if one does not pitch in to advance one's profession's prestige, power, and pay. I have a friend who is an airline pilot who flies jumbo jets. At the moment, his union is threatening to strike for a pay increase. He personally thinks that pilots are already overpaid and is free to say that and vote against the strike in union meetings. But if the motion to strike carries, he will be out there on the picket line, waiving his striker's placard. It is this—group dynamics, if you will—not the arrogance of individuals, that explains why science, which now holds the cards, "has shown itself unable to exist with anything." There is no institution today that has the power to say to science, *Stand back; that's my turf you're poaching on.*

I can remember the exact moment when this important fact broke over me like an epiphany. It was a decade or so ago and I was leading an all-day seminar on scientism in Ojai, California. As the day progressed, I found myself becoming increasingly aware of a relatively young man in the audience who seemed to be taking in every word I said without saying a word himself. True to form, when the seminar ended in the late afternoon, he held back until others had tendered their goodbyes, whereupon he asked if I would like to join him for a walk. The weather was beautiful and we had been sitting all day, but it was primarily because I had grown curious about the man that I readily accepted his invitation.

He turned out to be a professor at the University of Minnesota whose job was teaching science to nonscientists. Word of my seminar had crossed his desk, and being invested in the topic, he had flown out for the weekend. "You handled the subject well today," he said, after we had put preliminaries behind us, "but there's one thing about scientism that you still don't see. Huston, science *is* scientism."

At first that sounded odd to me, for I had devoted the entire day to distinguishing the two as sharply as I could. Quickly, though, I saw his point. I had been speaking *de jure* and completely omitting the *de facto* side of the story. In principle it is easy to distinguish science from scientism. All the while, in practice—in the way scientism works itself out in our society—the separation is impossible. Science's *conatus* inevitably enters the picture, as it does in every institution. The American Medical Association is an obvious example, but the signs are everywhere.

Jürgen Habermas, a philosopher of the Frankfurt School, coined a useful phrase for the way money, power, and technology have adversely affected the conditions of communication in ordinary, face-to-face life. He charged them with "colonizing the life world." A neo-Marxist himself, he had no particular interest in religion, but the concerns of this book prompt me to add scientism to his list of imperialists. One of the subtlest, most subversive ways it proceeds is by paying lip service to religion while demoting it. An instance of this is Stephen Jay Gould's book *Rocks of Ages*, which I will approach by way of a flashback to Lyndon Johnson. It is reported that when a certain congressman did something President Johnson considered reprehensible, Johnson called him into his office and said, "First I'm going to preach you a nice little sermon on how that's not the way to behave. And then I'm going to ruin you."

My nice little sermon to Professor Gould is, "Paleontologist though you are, you show yourself unable to distinguish rocks from pebbles, for a pebble is what you reduce religion to." Now for the ruination.

Of Rocks and Pebbles

Gould says he cannot see what all the fuss is about, for (he tells us) "the conflict between science and religion exists only in people's minds, not in the logic or proper utility of these entirely different, and equally vital subjects." When tangle and confusion are cleared away, he says, "a blessedly simple and entirely conventional resolution emerges," which turns out (not surprisingly) to be his own. "Science tries to document the factual character of the natural world, and to develop theories that coordinate and explain these facts. Religion, on the other hand, operates in the equally important, but utterly different, realm of human purposes, meanings, and values."

Note that it is human (not divine) purposes, meanings, and values that Gould's "religion" deals with, but the deeper issue is who (in Gould's dichotomy) is to deal with the factual character of the nonnatural, supernatural world. No one—for to his skeptical eyes the natural world is all there is, so facts pertain there only. He has a perfect right to that opinion, of course, but to base his definitions of science and religion on it prejudices their relationship from square one. For it cannot be said too often that the issue between science and religion is not between facts and values. That issue enters, but derivatively. The fundamental issue is about facts, period—the entire panoply of facts as gestalted by worldviews. Specifically here, it is about the standing of values in the objective world, the world that is there whether human beings exist or not. Are values as deeply ingrained in that world as are its natural laws, or are they added to it as epiphenomenal gloss when life enters the picture?

That this *is* the real issue is lost on Stephen Jay Gould, but not on all biologists. Two years ago I was asked to speak to the evolution issue at the University of California, Davis, in a lecture that its office of religious affairs arranged. Several days after returning home I received a letter from the biology professor who teaches the evolution course on that campus. He said that he had come to my lecture expecting to hear things he would need to refute at his next class session but had been pleased to find little of that nature in what I had said. Enclosed with his letter was an article he had written in which he raised the question of what the evolutionary fuss was about. His answer was: "It is not about whether or not evolution is good science, whether evolution or creation is a better scientific explanation of the diversity of life, or whether natural selection is a circular argument. The fuss actually isn't even really about *biology*. It is basically about worldviews." *Rocks of Ages* could have been a helpful book if Gould had recognized this point, but now, having had my fun with Gould, I must admit that I have not been entirely fair to him. For he is quite right in saying that the position he advocates is "entirely conventional." That does not make it right, but it does exonerate Gould from having invented the mistake, which I quoted Appleyard as indicating a few pages back. "Separating our values from our knowledge of the world [is the chief way scientism] burns away ancient authorities and traditions."

From Warfare to Dialogue

Religious triumphalism died a century or two ago, and its scientistic counterpart seems now to be following suit. Here and there diehards turn up—Richard Dawkins, who likens belief in God to belief in fairies, and Daniel Dennett, with his claim that John Locke's belief that mind must precede matter was born of the kind of conceptual paralysis that is now as obsolete as the quill pen—but these echoes of Julian Huxley's pronouncement around mid-century that "it will soon be as impossible for an intelligent or educated man or woman to believe in god as it is now to believe that the earth is flat" are now pretty much recognized as polemical bluster. It seems clear that both science and religion are here to stay. E. O. Wilson would be as pleased as anyone to see religion fail the Darwinian test, but he admits that we seem to have a religious gene in us and he sees no way of getting rid of it. "Skeptics continue to nourish the belief that science and learning will banish religion," he writes, "but this notion has never seemed so futile as today."

With both of these forces as permanent fixtures in history, the obvious question is how they are to get along. Alfred North Whitehead was of the opinion that, more than on any other single factor, the future of humanity depends on the way these two most powerful forces in history settle into relationship with each other, and their interface is being addressed today with a zeal that has not been seen since modern science arose.

This could be in part because money has entered the picture (the Templeton Prize for Progress in Religion is larger than the Nobel Prizes), but it probably signals a change in our climate of opinion as well. Scientists probably sense that they can no longer assume that the public will accept their pronouncements on broad issues unquestioningly, and this requires that they present reasons. In any case, God-and-science talk seems to be everywhere. Ten centers devoted to the study of science and religion are thriving in the United States, and together they mount an expanding array of conferences, lectures, and workshops. Several hundred science-and-religion courses are taught each year in colleges and universities around the country, where a decade or two ago you would have had to dig in hard scrabble to find one; and every year or so new journals with titles such as *Science and Spirit, Theology and Science,* and *Origins and Design* join the long-standing *Zygon* to augment the avalanche of books—many of them best-sellers—that keep the dialogue between science and religion surging forward.

On the whole, this mounting interest is a healthy sign, but it hides the danger that science (I reify for simplicity's sake) will use dialogue as a Trojan horse by which to enter religion's central citadel, which is theology. That metaphor fails, however, because it carries connotations of intentional design. A hole in a dyke serves better. If a hole appears in a Netherlands dyke, no finger in the dyke is going to withstand the weight of the ocean that pushes to enter.

Colonizing Theology

To once have belonged to the enemy camp provides one with insights into its workings, and so (with apologies for the military language) I will claim that advantage here.

When I came to America from the mission field of China, my theological landing pad at Central Methodist College in Missouri was naturalistic theism, the view that God must be a part of nature, for nature is all there is. With modest help from John Dewey, Henry Nelson Wieman was the founder of that school of theology, and my college mentor was one of his two foremost protégés. Thus it was that when I arrived at the Divinity School of the University of

Chicago to study with Professor Wieman, I was already as ardent a disciple as he had ever had. That lasted through my graduate studies, after which my resonance to the mystics converted me to their worldview.

At the time I am referring to (the middle of the twentieth century), Wieman's liberal naturalistic theism was giving its conservative rival—neo-orthodoxy, as founded by the Swiss theologian Karl Barth and captained in America by Reinhold Niebuhr—a run for the Protestant mind. Niebuhr won that round, but with Whitehead and his theological heir, Charles Hartshorne, naturalism has returned as Process Theology. Its philosophy of organism (as Whitehead referred to his metaphysics) is richer than Wieman's naturalism, and Whitehead's and Hartshorne's religious sensibilities were more finely honed, but Process Theology remains naturalistic. Its God is not an exception to principles that order this world, but their chief exemplar. God is not outside time as its Creator, but within it. And God is not omnipotent, but like everything in this world is limited. "God the semicompetent" is the way Annie Dillard speaks of this God.

Do we not see the hand of science—which process theologians point to proudly—in this half-century theological drift? In relating it to the concerns of this chapter, two questions arise. First, if we could have our way, would we prefer God to be fully competent or partially competent? Second, has science discovered any *facts* that make the first (traditional) alternative less reasonable than the second? If it has, science has vectored the drift and we must follow its lead. If no such facts have turned up, scientistic styles of thought are guilty of colonizing theology.

With this quick reference to the last fifty years, I turn now to the present.

The Tilt of the Negotiating Table

Because scientists at this point are negotiating from strength and would be happy to have things remain as they are, it is theologians who must take the initiative to get conversations going. I have already mentioned the ten or so religiously based institutes that are working at this job, and in these pages I shall confine myself to the two most prestigious of these, the Zygon Center at the University of Chicago, and the Center for Theology and the Natural Sciences at the Graduate Theological Union in Berkeley. In an informal division of labor, the Institute in Chicago publishes *Zygon*, the academic journal in the field, and the Berkeley Center mounts the conferences.

Who gets published in *Zygon* and invited to CTNS conferences? There is no stated policy; but an inductive scan suggests a bias against those who, first, criticize Darwinism; second, argue that the universe is intelligently designed; and third, accept the possibility that God may at times intervene in history in ways other than through the laws by which nature works. God may be believed to have created the universe and to operate within it, but God must not be taken to suspend at times its laws or to leave gaps in them that are divinely filled from outside. (That would give us a "God of the gaps," a deity who would be squeezed out when, as it is assumed will happen, science eventually fills those gaps.) In a word, miracles and supernaturalism generally are out. Those who honor the three mentioned proscriptions are welcomed in CTNS/*Zygon* doings; others are not.

Such at least is my reading of the matter. If the reading is basically accurate, the operative policy is pretty peculiar once one thinks about it. Three planks of the traditional religious platform have been removed by the pace-setting Berkeley/Chicago axis. (The religious platform I posit here is drawn from Hinduism and the Abrahamic religions. Buddhism and East Asia present complications that would be distractions in this discussion.) Why? The obvious answer seems to be that these planks do not fit the scientific worldview. I cannot speak for the governing boards of the two institutions and do not know if their policy here is tactical—to keep scientists from walking away from the negotiating table—or if it reflects a belief that science has discovered things that require that the traditional planks be dropped. I know the Berkeley team well enough to know that its members are sincere Christians who do not see themselves as capitulating to the scientific worldview if it is read in ways that exclude God. But the God they argue for is (1) the world's first and final cause, who (2) works in history by controlling the way particles jump in the indeterminacy that physicists allow them. This retains God, but in ways that supplement the scientific worldview without ruffling it.

The problem with this approach is that it overlooks the ghost of Laplace, who waits in the wings to announce that he has no need of the God-hypothesis. More serious is the procedural way things are going. The institutions that dominate the science-religion conversation do not consider the way they relate theology to science to be one possibility among others that merit hearings. They consider it to be the truth and believe that it needs to be understood if religion is to survive in an age of science.

Darwinism provides the clearest example of this monopolistic approach. That the issue of how we human beings got here has strong religious overtones goes without saying, and its founder and I are only two among millions who find the Darwinian theory (when taken to be fully explanatory of human origins) pulling against the theistic hypothesis. Among scientists themselves, debates over Darwin rage furiously, fueled by comments such as Fred Hoyle's now-famous assertion that the chance of natural selection's producing even an enzyme is on the order of a tornado's roaring through a junkyard and coming up with a Boeing 747. But when religion enters the picture, scientists close ranks in supporting Darwinism, with CTNS and *Zygon* right in there with them. To my knowledge, no one critical of the theory has been published in *Zygon* or been included in a major CTNS function.

Michael Ruse of the University of Guelph—a self-confessed bulldog for Darwinism—puts this colonization of theology by biology in perspective when he charges his fellow Darwinists with behaving as if Darwinism were a religion. Rustum Roy, a materials scientist at Pennsylvania State University, goes further. Half seriously, he has threatened to sue the National Science Foundation for violating the separation of church and state in funding branches of science that have turned themselves into religions. If these spokespeople are right and Darwinism has grown doctrinal, we have the curious spectacle of its colonizing not only theology but biology as well. I will close this chapter with an instance.

The 1999 conference on "The Origin of Animal Body Plans and the Fossil Record" was held in China because that is where a disproportionate number of fossils relating to the Cambrian explosion of phyla have been found. On the whole, its Western delegates argued that the explosion can be explained through a Darwinian approach, whereas the Chinese delegates were more skeptical of that. Jonathan Wells, of the Center for Renewal of Science and Culture at the Discovery Institute in Seattle, closed his report of the

conference with an account that carries overtones ominous enough
to warrant its being quoted in full:

> I will end this report with one poignant anecdote about a conver-
> sation I had with a Chinese developmental biologist from
> Shanghai who recently returned from doing research in Germany.
> She told me that in China the general practice in education is to
> settle on an official theory and then teach it to the exclusion of all
> others. So far, she said, this has not happened in biology; since she
> herself is a critic of the idea that genetic programs control devel-
> opment, she dreads the possibility of being forced to teach the
> Darwinian line. But she fears that this may happen soon, and she
> and her colleagues believe their only hope is the willingness of
> western scientists to discuss competing theories and not descend
> into dogmatism. It depressed her to see at this conference how
> dogmatic American biologists had already become, and she
> pleaded with me to defend the spirit of free inquiry. The way she
> put it, the world is counting on you to do this.

14

Life as Non-Historical Reality

Giuseppe Sermonti

22 The Lord possessed me in the beginning of his way,
  before his works of old.

23 I was set up from everlasting, from the beginning,
  or ever the earth was.

27 When he prepared the heavens, I was there:
  when he set a compass upon the face of the depth.

Proverbs, 8

By *historical* is meant not just any succession of events but a succession of such character that what follows implies (is derived from) what precedes it. Events need not only be serial, but their sequence must be such as to proceed in a single direction, the direction of history. A *catastrophe* is not historical: it is an abrupt occurrence not referable to an immediately preceding cause. The development of the embryo is a historical process, like the life of a man or of a people. The expanding Universe is a historical process (to which a stationary theory was opposed) but the rotation of the Earth and the revolution of the Earth around the Sun are not. It is not possible to distinguish one day from another or one year from another on simply astronomic grounds. That which is perfectly cyclic is not historical. Sea waves have no age. Likewise Biblical Wisdom (*Proverbs* 8:22-27) is not historical; it is permanent and forever. To what extent Life as a general phenomenon may be considered historical is the object of the present article, although surely its single expressions not only have historical features but, like birth, development and death, symbolize history itself.

The origin of life, the settlement of its biochemical composition or of its genetic structure, the formation of the various *taxa* are by an increasing number of scientists thought to have occurred *very early or very quickly*. In other words their historical process is deferred to the primordial stage and is excluded from recordable time. A stationary, balanced, cyclic situation exists thereafter, in which all historical features are lost. This emerging view opposes the evolutionary view according to which Life as a general phenomenon is a progressive process; it is continuously innovated and developed, and its structure is the result of a cumulative trend.

Some aspects of the living world will be discussed in this respect, leading to the eventual conclusion that a stationary (steady state) view accounts better for the observed facts than an evolutionary (historical) view. The problem of *origins* is outside the domain of our understanding from the scientific point of view.

The Constancy of DNA. The amount of DNA per nucleus in different organisms is divided into two orders of magnitude: millions of nucleoticle pairs in prokaryotes, and billions of nucleotide pairs in eukaryotes.[1] Intermediate values such as those of some moulds or insects do not figure as transitional. This difference in quantity reflects a profound difference in the DNA organization. The prokaryotic DNA is not protein-bound, exhibits no high repetitivity, does not have spacers between or within the genes (introns); all features which are present in the eukaryotic DNA. This structural difference leads to the question of whether the larger amount of DNA per nucleus in eukaryotes corresponds to an increase in information. In a recent paper by Orgel and Crick[2] the bulk of eukaryotic DNA is considered as junk or garbage. Its presence is attributed to a tendency toward an uncontrolled self-reproduction process, to such an extent that the larger part of DNA is defined as the ultimate parasite. It is present in the cell only because it is not harmful enough to warrant elimination. In a twin paper, Doolittle and Sapienza[3] question the "phenotype paradigm," *i.e.* the belief that DNA needs a means of expression and consequently a control by

1 A. H. Sparrow, H. J. Price and A. G. Underbrink, "A survey of DNA content per cell and per chromosome of prokaryotic and eukaryotic organisms: some evolutionary considerations," in: *Evolution of Genetic Systems*, ed. by H. H. Smith (New York: Gordon & Breach, 1972).

2 L. E. Orgel and F. H. C. Crick, "Selfish DNA: the ultimate parasite," *Nature*, 284 (1980), pp. 604-607.

3 W. F. Doolittle and C. Sapienza, "Selfish genes, the phenotype paradigm and genome evolution," *Nature*, 284 (1980), pp. 601-603.

the phenotype. In a previous paper, Doolittle[4] regards the eukaryotic-like DNA as the original primitive form of DNA. It would not have acquired but only "maintained the genetic plasticity present in the genomes of the ancestor common to all cells."

It is now widely accepted that the amount of DNA in various *taxa* of eukaryotes has no relation to the complexity of their phenotypic organization. Various authors who have collected data on the quantities of DNA per haploid nucleus (e.g. Britten and Davidson,[5] 1971; Sparrow *et al.*,[6] 1972) give for the Echinoderms values from 1 to 2 billions of nucleotide pairs (n.p.), for Anellids, 2.5; for Mollusca from 1.2 to 10; for Bony fishes from 0.35 to 5; for Reptiles from 3 to 6; for Mammals from 3 to 12; for Birds from 1.5 to 3. Values of 100 x 10^9 n.p. and more are reported for some Urodels and Dipnoals.

Thus the consideration of the amount of DNA as a direction in the history of Life, which some researchers had enthusiastically accepted is no longer valid. We may conclude: there has not been any evolution in the amount of DNA. In this respect, Man (6 x 10^9 n.p.) could just as well have appeared in the Cambrian era; together with Mollusca and Protozoa.

The Amount of Genetic Information. The amount of DNA eventually decoded in an organism can only be evaluated approximately. It can be deduced from the number of proteins which in turn is estimated, conservatively from the number of enzymes or functions. In prokaryotes, if one takes 300 as the average number of amino acid residues for each protein (900 nucleotide pairs) and 10% as the fraction of non-decoded DNA, one can assume that *one gene = c. one thousand nucleotide pairs.*[7] By this criterion *Escherichia coli* K 12 would be assigned c. 3,200 genes and the average bacterium about 5,000 genes.[8]

---

4 W. F. Doolittle, "Genes in pieces: were they ever together?" *Nature*, 272 (1978), pp. 581-582.

5 For example R. J. Britten and F. H. Davidson, "Repetitive and non-repetitive DNA sequences and a speculation on the origins of evolutionary novelty," *Quarterly Review of Biology*, 46 (1971), pp. 111-138; and Sparrow *et al.*, *ibid.*

6 *Ibid.*

7 J. D. Watson, *Molecular Biology of the Gene*, 3rd edition (New York: W. A. Benjamin Inc., 1976).

8 Sparrow *et al.*, *ibid.*

Another estimate of the number of genes, as deduced from the number of functions, was reported for the fruit-fly by Judd *et al.*[9] They have genetically analyzed with extreme accuracy some chromosomal regions in the salivary glands of *Drosophila* arriving at the statement: *one band = one function (one gene)*. The total number of bands in *Drosophila melanogaster* is c. 5,000. The estimate of 5,000 genes in an insect was quite surprising. In an organism so morphologically complex in comparison to a bacterium, the gene number would appear to be very similar to that of prokaryotes. Thus an insect does not require significantly more functions than a bacterium.

Based on a completely different principle, the number of functions has been estimated in some Amphibia such as *Xenopus* (African toad) and *Triturus* (newt). The evaluation is based on the observation of the number of loops surrounding the lump-brush chromosomes in the oocytes. Observation under electron microscope of RNA produced along a loop provides images similar to the so-called "Christmas tree" observed by Miller and Beatty[10] on bacterial or eukaryotic ribosomal DNA, and corresponding to a single transcript (gene). The equation *one loop = one gene* appears acceptable. Both *Triturus* and *Xenopus* (although the first has tenfold the DNA per nucleus than the latter) exhibit 5,000 loops per haploid genome. This number may well be only a rough approximation, and values at variance have been observed in other orders of Amphibia (but on the same order of magnitude). The general impression, however, is that the gene number is essentially unaltered in a bacterium, an insect, or a vertebrate. There are not in fact major differences between molecular types of uni- and multi-cellular organisms.[11]

Waddington and Lewontin[12] have proposed the so-called Serbelloni Theorem which states that "Every tendency to increase

9 B. H. Judd, M. W. Shen and T. C. Kaufman, "The anatomy and function of a segment of the X chromosome of *Drosophila melanogaster*," *Genetics*, 71 (1972), pp. 139-156.

10 O. L. Miller Jr. and B. R. Beatty, "Portrait of a gene," *Journal of Cellular Physiology*, 74 (1969), pp. 225-232.

11 F. Jacob, "Evolution and tinkering," *Science*, 196 (1977), pp. 1161-1166.

12 C. H. Waddington and R. C. Lewontin, "A note on evolution and change in the quantity of genetic information," in *Towards a Theoretical Biology*, ed. by C. H. Waddington (Edinburgh: Edinburgh University Press, 1968).

the quality of information in the genome will be held under control because the rate of progress under natural selection will be inversely proportional to the number of information units." The maximum number of genes could well have been reached *very early*.

Further studies obviously are required to reach a more reliable estimate of the number of genes in the various *taxa*; but we consider it a reasonable hypothesis that such a number is substantially invariant and remains around 5,000. This figure refers only to structural genes, "regulatory" genes being so far undetected in eukaryotes. A similar hypothesis, based on the number of protein species, was put forward by Omodeo.[13] "The cell of a fungus"—he wrote— "does not contain in all likelihood more protein species than a bacterial cell . . . thus a protozoan wouldn't have many more proteins than a fungus. As far as enzymes are concerned, these are produced in even fewer number. The same holds for Metazoa."

The picture emerging from these considerations is that during the terrestrial presence of life there has not been a progressive modification of the structural material in which the genetic information is memorized or coded, but rather a variation (not necessarily adaptive) in the phenotypes, whose genetic memorization (assimilation) was a secondary effect and, therefore, not a primary cause of variation.

Origin of Biochemical Differentiation. The diversity, unapparent in the number of genes, could be revealed in the quality of the genes, i.e. in their structure and function in the various *taxa*. Study of the primary structure and function of numerous ubiquitary proteins has produced surprising results.[14] This is a well known chapter. It can be summarized by stating that the greater the distance between two species, the higher the number of amino-acidic residuals in cytochrome C by which they differ. This was expected. The astounding part of the story is that the spatial configuration and the function of all cytochromes C so far examined, from man to reptiles, to fishes, to flies, to moulds, all are superimposable. The differences are not *adaptive* and concern regions with no relevance to

---

13 P. Omodeo, *"Evoluzione del genoma alla luce dell'informatica,"* Atti Assoc, Genet. Ital., 21 (1976), p. 166.

14 E. Margoliash, W. M. Fitch and R. E. Dickerson, "Structure, function and evolution of proteins," *Brookhaven Symposium of Biology*, 21 (1968), p. 259.

the function. Natural selection has played a conservative role in maintaining intact that part of the gene corresponding to the amino acids involved in the function.[15] In other words, the diversification among proteins is determined through *neutral* mutations.[16] The same holds for proteins such as fibrinogen, globins, proinsulin, histone IV A. The latter undergoes the substitution of a residual every 200 million years. However, when the gene for histone IV A of two sea urchins was (partially) deciphered, out of 27 third-position nucleotides, as many as 9 were different, and the affected triplets were synonymous. A first-position difference was also synonymous and relative to a six-codon amino acid (leu).[17] This shows that mutation is also active in the apparently stable genes, and the reduced variability is the result of a strong functional constraint which causes the loss of a large number of mutants. What appears to be clear from this research is the fact that it is not the variation in DNA which has produced the differentiation among *taxa* at the gene level, but rather the separation of *taxa* has permitted the (neutral) diversification among the genes by interrupting the genetic flow between separating groups. Neutral modifications are not historical: they do not define a direction in transformation (as the adaptive ones do).

As a result of this essential gene constancy, the cell metabolism remains substantially uniform in all organisms. It is the latter which has experienced the pressure of natural selection, transferring it to the genes. And what we have stated for genes can also be said of biochemistry. "It is not biochemical novelties which have generated diversification of organisms," wrote F. Jacob in 1977. "In all likelihood, it worked the other way around. . . . What distinguishes a butterfly from a lion, a hen from a fly, or a worm from a whale is much less a difference in chemical constituents than in the organization and distribution of these constituents. Biochemical uniformity is preserved by the genetic flow within species and—to a less restric-

15 R. E. Dickerson, "The structure of cytochrome C and the rates of molecular evolution," *Journal of Molecular Evolution*, 1 (1971), pp. 26-45.

16 M. Kimura and T. Ohta, *Theoretical Aspects of Population Genetics* (Princeton, N.J., 1971.

17 M. Grunstein, P. Schedl and L. Kedes, "Sequence analysis and evolution of sea urchin histone H4 messenger RNA," *Journal of Molecular Biology*, 104 (1976), pp. 351-369.

tive extent—by the constraint of natural selection. When the former is interrupted, biochemical differences can appear in metabolic reactions no longer under the control of natural selection. They are not necessarily adaptive, and as far as we know, are rather losses. On the whole, all biochemical variation in the biosphere is marginal or largely neutral. Biochemical changes do not seem," Jacob[18] wrote, "to be a main driving force in the diversification of living organisms. The really creative part in biochemistry must have occurred *very early*." (my italics)

Quantum Leaps among Fossils. Among the records of paleontology, the part relevant to our argument is that which distinguishes the progressive from the steady-state picture. A succession of abrupt appearances does not differ from a steady-state situation, with the starting points scattered along the time.

The gradualness is first to be questioned at the very origin of life. That life was shaped on the Earth is more and more doubtful. The oldest "compelling" evidence of life was considered to come from the superbe stromatolites in Canada (2.7×10^9 years), while the oldest "possible" from Isua in Greenland (3.7×10^9 years).[19] Recent reports of bacterial chains from the "North Pole" of Australia antedate the first "compelling" evidence for life to 3.5×10^9 years ago. "If life did originate on Earth," E. G. Nisbet wrote, "the processes leading up to it must have happened *very quickly indeed*" (my italics). The age of the oldest traces of life approach that of the oldest rocks (3.8-3.9×10^9 years). These figures support the hypothesis of life coming to the Earth from outer space (on a meteorite or in the tail of a comet?) and shift to the Infinite the problem of the Origin.[20]

The "abiogenetic" theories are thus losing support. Even more so if we accept that the organisms of the "North Pole" were most likely photosynthetic and that the presumed methane/ammonium atmosphere of the primeval Earth should no longer be given serious consideration.

---

18 Jacob, *Ibid.*

19 E. G. Nisbet, "Archean stromatolites and the search for the earliest life," *Nature*, 284 (1980), pp. 395-396.

20 F. Hoyle and C. Wickramasinghe, *Lifecloud* (Mondadori, Milano: Trad. ital. *La nuvola della vita*, 1979).

The explosive appearance of all the main *phyla* of Metazoa at the beginning of the Cambrian age (600 million years ago) is firmly established (only the Chordata would have appeared in the successive Ordovician period). No animal *phylum* came forth in the following epoch, not even when Metazoa colonized the dry land. This may be described as the (catastrophic) conformation of living matter, as soon as the cellular state was achieved, to conform to the entire possible series of structural models available to metazoic growth. The gradual and haphazard development (by mutation-selection) of the types through adaptation to the changing environmental conditions, according to the historical evolutionary theory, would have produced the diversification of the large *phyla* as a final result and not as a first.[21]

The explosive "radiation" of *taxa*, with all their subdivisions and the virtual absence of intermediate links, is the rule in paleontology.[22] The best known example is in the mammals. All the orders appeared in a geologically short period of time and already perfectly formed, out of a form (*the mother*) which, according to Grasse[23] could not have been a specialized reptile, but rather a such primitive being to be almost identified with the mother of reptiles. Mammals would thus not be born from reptiles, but among reptiles. The "experience" of reptiles could not be transferred to mammals.

The opinion that *taxa* appeared abruptly, by a kind of *quantum leap*, not necessarily in a direction that represents an obvious improvement in fitness, is gaining increasing credit. This view of "punctuated equilibria" cannot be said to be evolutionary. Genealogical trees, transitional forms and progressive adaptations, which are the factual basis of Darwinism are not consequential to it.

Was the diversification of life progressive? Did the major *taxa*, as the stratigraphic succession suggests, appear first and then, little by little, the minor? We have been particularly impressed by the arguments of David Raup who has listed as many as seven factors (the main being the antiquity of the beds) which would conceal a substantially stable state of diversification in the biosphere. On the

21 G. Sermonti and R. Fondi, *Dopo Darwin critica all'evoluzionismo* (Milano: Rusconi, 1980).

22 P. P. Grasse, *L'Evolution du vivant* (Paris: Albin Michel, 1979).

23 *Ibid.*

basis of complex statistical work, Raup and Stanley[24] conclude that the present diversification (the assumed 4.5 million species) might not be significantly different from that in the Cambrian era or later. According to this view, the biosphere underwent transformation but not evolution, at least as far as the diversification of living beings is concerned. As repeatedly stated by Grasse,[25] "Transformation is not evolution."

The Geometric Constants. The work by D'Arcy Wentworth Thompson[26] *On Growth and Form* (first edition 1917) is more and more frequently quoted in scientific writing. His central idea is that nature is simply a reflection of the forms conceived in geometry. Form problems are essentially mathematical, and growth problems are essentially physical. Morphogenetic solutions are primarily the result of a geometric pattern of growth, and secondarily an adaptation to the constraints of natural selection. The recognition of a "natural law of structure in the taxonomic system,"[27] represents the affirmation of a universal harmony at the basis of *systema naturae.*

At the microscopic level, the presence of the Platonic solids and of deltahedrons in the forms of Radiolars (and viruses) is the most striking example of an indispensable structure in nature. The geometrical necessity of phyllotaxis according to Fibonacci's series (each term is the sum of the two preceding it)

$$1, 1, 2, 3, 5, 8, 13, 21, 34 \ldots$$

was recently restated by Mitchison.[28]

The special relevance of the geometric view lies in the fact that the physico-mathematical rules which it implies are not historical. This is clearly stated by D'Arcy Thompson: "*In the physico-mathematical order of complexity, sequence and historic time are out of the question.*" (my italics)

24 D. Raup and S. Stanley, *Principles of Paleontology* (San Francisco: Freeman, 1971).

25 Grasse, *Ibid.*

26 D'Arcy W. Thompson, *On Growth and Form,* ed. by J. T. Bonner (abridged ed.) (Cambridge: Cambridge University Press, 1961).

27 A. Sacchetti, *Rivista di Biologia,* 74 (1981), p. 1.

28 G. J. Mitchison, "Phyllotaxis and Fibonacci series," *Science,* 196 (1977), pp. 270-275.

The thinking of D'Arcy Thompson was adopted and deepened by René Thom[29] who formulated a theory of morphogenesis *in abstracto*, purely geometrical, independent of the substrate of the forms and the nature of the forces which create them. Developed to the point of paradox, this theory does not explain why there is not but a single form. Thom[30] attributes the cause of morphogenetic differences to what he calls *elementary catastrophes*, determined by the topological structure of the internal dynamics. The morphogenetic laws, functioning as the form builder, release biological information from the task of shaping forms, leaving to it the more modest function of opting for one or another *structurally stable* model.

Related considerations are to be found in the late Waddington,[31] assessing the canalized development of the epigenetic trajectories (creods) or the principle of archetypes. Some living forms appear as an inevitable realization of some morphologic typologies, the necessity of which is illustrated by Waddington through geometric metaphors. It is by now an obsolete concept that Nature could have arrived at any form whatever and that natural selection would have chosen the most compatible with survival. This Empedoclean idea was abandoned by Darwin[32] himself who in the *Descent of Man* wrote that "in most cases we can only say that the cause of any small variation and of any monstrosity is more in the nature and constitution of the organism than in the nature of the surrounding conditions."

Since the beginning, Life has had an essentially constant genetic-biochemical structure. Its morphological variability is moreover under the control of physico-mathematical constants also invariant in time. In both regards—the complexity present from the beginning and the geometrical rules present (as Wisdom) outside time—Life is not historical.

Historical (evolutionary) processes in the realm of Life are likely confined to some so-called orthogenetic phenomena, the nature of which is still elusive, aside from some degenerative phenomena, more easily to be figured in connection with the law of entropy.

29 R. Thom, *Stabilité structurelle et morphogénèse. Essai d'une théorie générale des modèles* (Paris : Inter editions, 1972).

30 Thom, *Ibid.*

31 C. H. Waddington, *The Evolution of an Evolutionist* (Edinburgh: Edinburgh University Press, 1975).

32 C. Darwin, *Descent of Man.* Vol. I (1869), p. 152.

15

Man, Creation and the Fossil Record

Michael Robert Negus

The Theory of Evolution is a biological paradigm proposed by western civilization that is contra-traditional and tends towards atheism. It is impossible that such a proposal could have arisen in a civilization that was centered upon the Spirit and guarded by Tradition. Evidence for the theory is derived from fossil remains, the gathering and accumulation of which is laborious, involving considerable excavation, bringing to light things which have, in the natural course of events been buried. The difficulty of obtaining such information may be counted as a blessing, since it is not only concerned with what is residual and accidental, but unlike the study of living things, does not readily open the mind to the transcendent.

However, that does not mean that the fossils do not have their first principle in God, far from it. In fact the function of this paper is to propose an interpretation of the fossil record within a broad framework of cycles and principles, rather than from a viewpoint of either "randomness" or "progress" which characterize profane ways of thinking.

Before trying to interpret the characteristics of previous cycles it is necessary to understand the cycle in which we live and the relationships within it. Because of the traditional analogy between microcosm and macrocosm, there is a relationship between the human constitution, especially the human psyche, and the collectivity of all other beings on earth.[1] Human consciousness is linked

---

1 It would seem that the analogy exists because of a kind of "resonance" that is maintained between the microcosm and the macrocosm. It is important to remember that each domain has its own integrity and detail, and so the correspondence should not be pursued exhaustively.

up with the individual spirit, the mental faculties (reason, memory, and imagination) and the interior continuities of the senses and organs of action. Consciousness is the uniting principle of these three broad divisions of the psyche. Likewise among non-human beings the vertebrate form is the uniting principle of the three vertebrate types, birds, mammals and fish (i.e. the bony fish, *Teleostei*) which characterize our cycle[2] and which correspond to the three psychic domains mentioned above. The relationship is particularly clear with reference to the forms and motion of these animals. Birds obviously manifest a spiritual or aerial nature, mammals by their complex movements in a two-dimensional plane, manifest an expansive one, whereas fish are "trapped" within an aquatic domain that corresponds to sensory experience. The invertebrate animals and plants correspond to the lower part of the psyche, closely linked to the body. The invertebrates represent the emotive and reflexive aspects, while the plants, the physiology of which is inseparably linked to the sun as well as the atmosphere, represent the vital part of the psyche that blends with the life of the body.

Sufficient has been said to indicate the harmony that exists between the external and internal worlds. But it should be remembered that the principle of harmony also operates within each world; the principle in fact being identical to the Hindu concept of *Dharma*. This Cosmic Harmony is the principle that accounts for the co-ordination of all the beings in the cosmos and which maintains, on the terrestrial plane, "ecological" integration. One of the consequences of harmony is the process of "natural selection," so cherished by evolutionists. In fact, natural selection is one aspect of the principle of natural harmony. It acts in two ways. Firstly, it maintains the norm of the species by eliminating serious deviations or mutations, which result from the imperfections of biological, reproductive processes. Secondly, it acts selectively upon the possibilities that result from the genetical variation of a species; this enables the development of close, "adaptive" adjustment between the various species in an ecosystem and between the species and their environ-

2 By "our cycle" one means the period of perhaps 100 million years ago to the present time, so including the upper part of the Cretaceous and the whole of the Cenozoic era. During this period the animals and plants which characterize our world had their origins and the world came to take on the nature and ambience it has today.

ment. The overall effect is to increase the manifestation of harmony. Adaptation is allowed by a degree of flexibility of the psychic and corporeal aspects of a being; however, there appear to be limits beyond which harmonious flexibility is impossible. The amount of harmonious adaptation that any one species is capable of depends on the nature of the primitive species, but undoubtedly in some cases the change is considerable. For example the ass, the horse and the zebra probably comprise one "species," all being derived from the primitive form which contained within itself the possibilities later manifested by the derived and isolated forms. Primitive forms are generally less specialized and more open to adaptive change. Later forms are typically specialized and isolated. Their form and physiology corresponds to restricted combinations of possibilities inherent in the primitive ancestor. Thus, we see that whereas adaptation enhances the harmony between species and the environment, it also means specialization and finally, disharmony and extinction.[3]

These cyclical events lead us to a consideration of the much greater cycles involving large numbers of beings and species which are illustrated by the so-called fossil record. Evolutionists have great difficulty in applying an evolutionary pattern to the invertebrate animals and to the plants, except for the adaptive sequences described above.[4] They focus mainly on the vertebrate animals and these will also be our chief concern.

---

3 *Editor's note:* "The traditionalist has no argument with the evolutionist so far as these facts are concerned. The evolutionist uses them as evidence for the Theory of Evolution; the traditionalist interprets them as illustrating the flexibility of a species, the means by which organisms are capable of optimal integration with one another and with their environment. To some extent the evolutionist would agree. However, in one respect the two points of view are completely opposed: the traditionalist regards change as implying some kind of loss, even though adaptive, whereas the evolutionist regards change as implying, in principle at least, some kind of progress." (From *Reactions to the Theory of Evolution* by Michael Negus).

4 The invertebrate phyla appeared suddenly and with great diversity in the Cambrian era some 550 million years ago. The lack of sequential pattern is because of the "substantial" nature of these peripheral beings, that is to say that they are only weakly involved in vertical "resonance" with the human microcosm. Those invertebrates that appeared at the beginning of our cycle, for example the butterflies and bees together with the flowering plants associated with them, do "resonate" essentially with the human microcosm, hence they manifest a strong spiritual symbolism.

The event of creation is described by traditional sources in different ways, depending upon the point of view. For example in the Koran: "His [Allah's] is the origin of the heavens and the earth. When He commands a thing He says to it: 'Be!,' and it is" (2:117). This is creation in principle. "He is Allah, the Creator (*Khâliq*), the Bringer-into-existence (*Bâri*), the One-who-gives-physical-form (*Musawwir*)" (59:24). This shows the continuity between the free will of the Creator, the willed act of creating and the transcendent cause of physical form and hence the cause of symbolism. In the Book of Genesis "the Lord God formed man from the dust of the ground, and breathed into his nostrils the breath of life; and man became a living soul (Hebrew *nephesh*)" (Genesis 2:7). This quotation shows the single outcome of two acts. Bodily formation is described first, then its animation that results in a single living being. One can also understand creation as continuous but consisting of an indefinite series of single moments, the renewal of creation at each moment (*tajdîd al-khalq bi'l-anfas* literally "each breath"), this being a Sufi doctrine.[5] Whatever the doctrine, creation always has two complementary dimensions, one vertical in conformity with the Will of God and one horizontal, in nature, a *creatio continua* which allows for the integrating and adaptive processes of change. Given the fact that such natural changes tend to an increase in harmony we could see them as an *attraction* towards the Creator. Only the latter movement takes place in time and space and consequently it is only this dimension which is represented in the fossil record.

The account of creation in the Book of Genesis provides a suitable guide for the gross interpretation of the fossil record.[6] Genesis describes six cycles beginning with the creation of the Intellectual Light (Sanskrit *Buddhi*) and ending with the creation of man. The relationship between the two does not have to be stressed. It is this Light that is responsible for the "resonance" and so the symbolism that makes it possible to compare the microcosm and macrocosm.

---

5 See Titus Burckhardt's *An Introduction to Sufi Doctrine* (Lahore, Ashraf, 1989), Chapter 10.

6 The symbolism of Genesis is complex and synthetic. The commands are not given in time, no more than each cycle is actually one day. The fossil record is a temporal, sequential reflection of the commands which are in themselves supra-temporal.

In the second cycle we see the separation of supra-formal and formal possibilities ("upper and lower waters") and in the third cycle the origin of the formal possibilities in the creation of "earth," that is the whole of mineral creation.[7] The procession of "plants," which ends the third cycle and inaugurates the fourth, coincides with the creation of the luminaries.[8] The fourth cycle during which the full photosynthetic activity of the green plants developed led to an immense change in the world. During that cycle the level of oxygen in the atmosphere increased to a level similar to that of today.[9] This produced the aerobic conditions necessary for animal and human life, which followed.

The cycles are developments of the possibilities contained synthetically in the Intellectual Light. The harmony of each cycle is due to the integrity of the Intellect and, so far as the formal world is concerned, to the Law that regulates each cycle, giving each its particular characteristics. Although each cycle is a creation in its own right, each must interact with the cycle that follows it so that by means of some adaptation and some elimination a greater and more complex harmony comes to exist.

The fifth and sixth cycles are of particular interest as far as this paper is concerned since they correspond to the formation of the greater part of the fossil record. The fifth cycle is begun by the creation of the "fish" and ended by the creation of the "birds," while the sixth cycle is begun by the creation of "land beasts" (i.e. mammals) and terminated by the creation of man. The duration of the fifth cycle corresponds to the vast geological period between the

---

7 In modern cosmology the term "earth" as used in Genesis, means the physical cosmos. Science ignores the symbolism of the "heavens," which for traditional peoples really was the transcendent realm. The creation of the "earth" described in Genesis would therefore correspond to the "singularity" that was the origin of all physical matter, probably about 15 billion years ago.

8 This sequence, that "plants" arise from the conditions of the cycle before that of the Sun, would suggest that the first "plants" were not actually dependent upon the Sun as a source of energy. Such organisms are referred to nowadays as *chemolithotrophs*. These microorganisms, like green plants, make use of carbon dioxide as their source of carbon.

9 The appearance of atmospheric oxygen, initially due to the photosynthesis of blue-green bacteria, probably began about 2.3 billion years ago. The gradual increase of oxygen to a balance of 21% is a nice example of the principle of harmony operating.

middle of the Old Palaeozoic and the end of the Mesozoic.[10] The first fossils to be found are fish,[11] then during the New Palaeozoic and the Mesozoic there are amphibians and reptiles. Finally, at the end of the Mesozoic fossil birds appear. Here we see a linear and successive emergence of possibilities between two poles, the lower corresponding to the palaeozoic fish and the upper to the cretaceous birds. We would expect to find a continuous series of possibilities between these two extremes; thus in addition to amphibians and reptiles we find fish amphibians, amphibian-reptiles and reptile-birds. The fifth cycle contains a closely packed "sweep" of possibilities making up a creative "movement" from aquatic, through terrestrial forms, finally to birds that ascend above the earth's surface. At the same time there are several secondary or minor cycles operating in parallel to and in conjunction with the main cycle. For example, during the Permian, Triassic and Lower Jurassic periods, in addition to the Cotylosaur/Thecodont reptiles which are "central" in relation to the ascending line from fish to birds, we see the creation of creatures such as the tortoises and the mammal-like reptiles. The latter beings are of considerable interest since they are creations of a secondary cycle that includes the non-eutherian mammals (e.g. the monotremes and marsupials). This secondary cycle forms at the interface of the major fifth and sixth cycles, maintaining a continuous link of possibilities between them.

The sixth cycle includes the creations of mammals and man. Man occupies a central position in creation since "made in the image of God" he is the only being which has direct access to the supra-formal world. It is significant that subsequent to the creation of man the Creator is said to "rest," that is to say He "retreats" to a Center of "actionless activity," which constitutes the inner reality of which man is the exterior aspect.

Man is not only the image of God, he is also a eutherian mammal, and these two parts of his nature are inseparably linked. Within the *Eutheria* two divisions can be seen: those animals that bear a close physical resemblance to man, namely the *Primates* and those that are diverse in structure and are unlike man. The latter

10 From about 550 million years ago to about 65 million years ago.

11 These are very ancient fish, not the "modern" bony fish that characterize our own cycle.

group includes all the non-primate *Eutheria*. The supreme form of man is seen in its "dominance" in the primate body, and also by the dominance of the primate characteristics over the early eutherians. In fact the first eutherians of the Cretaceous were intermediate between primates and insectivores, the insectivores being the most "peripheral" of the mammalian beings. The events of the sixth cycle have two characteristics: a) the successive approximation from "diffuseness" (earliest eutherians) to "centrality" (man); b) the creation of beings "intermediate" between the "periphery" (insectivores) and the "center" (man). The "intermediate" beings which include the rodents, carnivores, horse and cow types etc., need not concern us but the primates are of considerable interest because of their formal resemblance to man. The insectivores, pro-simians (e.g. lemurs, lorisoids, tarsiers), monkeys and apes make up a concentric series of "grades," resembling tree rings, arranged hierarchically around the center, man. The fossil record indicates the presence of a number of species that one would place within a grade between apes and man. They are a complex group and with respect to their skeletal forms are "man-like" to various degrees. They could be called "hominids" to distinguish them from man. They all belong to the macrocosm of the sixth cycle. Man may be regarded in one sense as the final point, the culmination of the sixth cycle. But in essence man belongs to the seventh cycle, when the Creator "rests" at the center of his creation, at the *axis mundi* of our world.

The earliest fossilized human skeletons are found in the Klasies River caves about 115,000 years ago ("Middle Stone Age"). These people are indistinguishable from modern humans (*Homo sapiens*).[12] It is likely that they resembled the Bushmen who are still found, albeit only residually, in Southern Africa.[13] They were hunter-gatherers and, although they are the oldest humans known,

12 Fragmentary evidence tells us that the Klasies River people fished and harvested marine shellfish, perhaps using boats, that they used a variety of stone tools, hunted the giant buffalo and they slept in woven grass mats (R. Singer and J. J. Wymer *The Middle Stone Age at Klasies River Mouth in South Africa*, University of Chicago Press 1982).

13 The book by Laurens van der Post and Jane Taylor entitled *Testament to the Bushmen* (Viking, 1984), includes an account of the spirituality of these remarkable people at what was perhaps the final days of their independence from modern influence.

were not of the first human population. Various methods have been employed to identify the first humans. These include molecular genetics[14] archeology[15] and linguistics.[16] This leads one to suggest a time for the origin of the human species, *Homo sapiens*, of approximately 150,000 years ago, with the location of that population in East Africa. It is thought that every human being in the world is derived by migration from this "founder" population.

The earliest human remains in Europe occur about 40 to 30 thousand years ago. These people belonged to the Cro-Magnon culture of the "Reindeer Age" and were responsible for the beautiful cave paintings in southern France and Spain. The evidence suggests that their culture and tradition were similar to recent North American native people. The evidence suggests that the earliest human traditions were based upon a Shamanistic mythology and upon the symbolism of nature.[17]

So far we have only considered the creative aspects of the cycles but in addition to these "positive" effects there are also "negative" or destructive effects. We shall consider two cases of destruction, both of which indicate "anticipation" of future events. The greatest example of destruction is that which terminated the fifth cycle, when all the archosaurian reptiles disappeared, never to be seen again. This catastrophic event coincided with the creation of the first birds and mammals. Earlier in the fifth cycle, when the first reptiles were being created, an aquatic destructive phase took place when almost all the Palaeozoic marine vertebrates were eliminated. The seas were then repopulated with the modern type of teleostean

14 This method studies mitochondrial DNA from many different world populations. The technique traces ancestral populations down maternal lines. This leads to one hypothetical ancestral mother, called "mitochondrial Eve."

15 Using sophisticated dating techniques to study first settlement dates, with particular reference to the age of hearths.

16 The tracing of word roots to "proto-languages." This technique does not have the quantitative value of the other methods, but it does closely follow genetical lines since both genes and languages divide as populations divide.

17 Readers may wonder how this information relates to the mythology of Adam and Eve in the Book of Genesis. The Genesis "history" of man conveys a story "out of time." What is described as a "place" (Paradise) is actually a state. The fossil remains of early human populations are very much in time and, in terms of Genesis, are "after the Fall," that is to say in a lower "earthly" state, without "closeness" to God.

fish later in the fifth cycle, just before the creation of the birds. The creation of the teleostean fish was the first sign of the approaching sixth cycle, which would bring about the elimination of the reptiles and their replacement by mammals. Such events indicate the complex and overlapping nature of the cycles and illustrate the fluidity of the creation process. The outcome for the sixth cycle was to produce the three prime vertebrate types for our own world, birds for the air, mammals for the surface of the earth and fish for the waters. This recalls the psychological considerations at the beginning of this paper.

The invertebrate animals and the plants constitute the lowest part of the macrocosm and consequently remain relatively passive to the cycles in comparison with the vertebrate animals. However, this "passivity" supports the characteristics of the world of the appropriate cycle. Thus, the *Pteridophyta*[18] (Ferns and Club Mosses) belong to the fifth cycle and the *Angiospermae* (Flowering Plants) to the sixth.

At the beginning of this paper it was said that there is an analogy between the unifying nature of human consciousness and the phyletic unity of the vertebrate form. Man is thus not only the central and principal mammal, but he is also the principle of the vertebrate body and psyche. This is clearly seen in embryology. It is a well-known fact that the young stages of different vertebrates resemble one another more than they resemble the adult stages, and more than the adult stages resemble one another. When the embryology of man is compared with that of other vertebrates, it is seen that the early stages are very similar. In later development, however, the non-human vertebrates "deviate" and become specialized in particular respects. For example the visceral pouches which occur in the pharyngeal region of all vertebrate embryos are arrested in development in fish to give rise to the gills, whereas in man their development continues until eventually they are converted into glandular organs and parts of the ear. Using the analogy of a tree we can say that whereas the embryology of man proceeds along the central axis or trunk, that of other vertebrates deviates in

18 The *Pteridophyta* (*e.g.* ferns, club-mosses and horsetails) were dominant during the fifth cycle. Today a limited number persist amongst the flowering plants (*Angiospermae*) that dominate our world.

various places giving rise to side branches, each of which may be regarded as a particular specialization of the trunk. Therefore the form of man is the vertebrate "norm" and the other vertebrate forms are aspects. This is a biological way of saying that Man is an expression of the Intellect and animals are particularizations of the Intellect.

This paper is in no way intended to be "dogmatic," but rather is written in the spirit of traditional science, in which the study of nature is pursued for the purpose of allowing students to discover themselves. It is also intended to show that traditional texts are not at all naïve but are exact and true. This is because of their spiritual origin, which is also the origin of the whole creation.

16

The Act of Creation: Bridging Transcendence and Immanence

William A. Dembski

Introduction

"Sing, O Goddess, the anger of Achilles son of Peleus, that brought countless ills upon the Achaeans." In these opening lines of the Iliad, Homer invokes the Muse. For Homer the act of creating poetry is a divine gift, one that derives from an otherworldly source and is not ultimately reducible to this world. This conception of human creativity as a divine gift pervaded the ancient world, and was also evident among the Hebrews. In Exodus, for instance, we read that God filled the two artisans Bezaleel and Aholiab with wisdom so that they might complete the work of the tabernacle.

The idea that creative activity is a divine gift has largely been lost these days. To ask a cognitive scientist, for instance, what made Mozart a creative genius is unlikely to issue in an appeal to God. If the cognitive scientist embraces neuropsychology, he may suggest that Mozart was blessed with a particularly fortunate collocation of neurons. If he prefers an information processing model of mentality, he may attribute Mozart's genius to some particularly effective computational modules. If he is taken with Skinner's behaviorism, he may attribute Mozart's genius to some particularly effective reinforcement schedules (perhaps imposed early in his life by his father Leopold). And no doubt, in all of these explanations the cognitive scientist will invoke Mozart's natural genetic endowment. In place of a divine afflatus, the modern cognitive scientist explains human creativity purely in terms of natural processes.

Who's right, the ancients or the moderns? My own view is that the ancients got it right. An act of creation is always a divine gift and cannot be reduced to purely naturalistic categories. To be sure, creative activity often involves the transformation of natural objects,

269

like the transformation of a slab of marble into Michelangelo's David. But even when confined to natural objects, creative activity is never naturalistic without remainder. The divine is always present at some level and indispensable.

Invoking the divine to explain an act of creation is, of course, wholly unacceptable to the ruling intellectual elite. Naturalism, the view that nature is the ultimate reality, has become the default position for all serious inquiry among our intellectual elite. From Biblical studies to law to education to science to the arts, inquiry is allowed to proceed only under the supposition that nature is the ultimate reality. Naturalism denies any divine element to the creative act. By contrast, the Christian tradition plainly asserts that God is the ultimate reality and that nature itself is a divine creative act. Within Christian theism, God is primary and fundamental whereas nature is secondary and derivative. Naturalism, by contrast, asserts that nature is primary and fundamental.

Theism and naturalism provide radically different perspectives on the act of creation. Within theism any act of creation is also a divine act. Within naturalism any act of creation emerges from a purely natural substrate—the very minds that create are, within naturalism, the result of a long evolutionary process that itself was not created. The aim of this article, then, is to present a general account of creation that is faithful to the Christian tradition, that resolutely rejects naturalism, and that engages contemporary developments in science and philosophy.

The Challenge of Naturalism

Why should anyone want to understand the act of creation naturalistically? Naturalism, after all, offers fewer resources than theism. Naturalism simply gives you nature. Theism gives you not only nature, but also God and anything outside of nature that God might have created. The ontology of theism is far richer than that of naturalism. Why, then, settle for less?

Naturalists do not see themselves as settling for less. Instead, they regard theism as saddled with a lot of extraneous entities that serve no useful function. The regulative principle of naturalism is Occam's razor. Occam's razor is a principle of parsimony that requires eliminating entities that perform no useful function. Using Occam's razor, naturalists attempt to slice away the superstitions of the past—and for naturalists the worst superstition of all is God. People used to invoke God to explain all sorts of things for which we now have perfectly good naturalistic explanations. Accordingly, God is a superstition that needs to be excised from our understanding of the world. The naturalists' dream is to invent a theory of everything that entirely eliminates the need for God (Stephen Hawking is a case in point).

Since naturalists are committed to eliminating God from every domain of inquiry, let us consider how successfully they have eliminated God from the act of creation. Even leaving aside the creation of the world and focusing solely on human acts of creation, do we find that naturalistic categories have fully explained human creativity? Occam's razor is all fine and well for removing stubble, but while we're at it let's make sure we don't lop off a nose or ear. With respect to human creativity, let's make sure that in eliminating God the naturalist isn't giving us a lobotomized account of human creativity. Einstein once remarked that everything should be made as simple as possible but not simpler. In eliminating God from the act of creation, the naturalist needs to make sure that nothing of fundamental importance has been lost. Not only has the naturalist failed to provide this assurance, but there is good reason to think that any account of the creative act that omits God is necessarily incomplete and defective.

What does naturalism have to say about human acts of creation? For the moment let's bracket the question of creativity and consider simply what it is for a human being to act. Humans are intelligent agents that act with intentions to accomplish certain ends. Although some acts by humans are creative, others are not. Georgia O'Keefe painting an iris is a creative act. Georgia O'Keefe flipping on a light switch is an act but not a creative act. For the moment, therefore, let us focus simply on human agency, leaving aside human creative agency.

How, then, does naturalism make sense of human agency? Although the naturalistic literature that attempts to account for human agency is vast, the naturalist's options are in fact quite limited. The naturalist's world is not a mind-first world. Intelligent agency is therefore in no sense prior to or independent of nature. Intelligent agency is neither *sui generis* nor basic. Intelligent agency is a derivative mode of causation that depends on underlying naturalistic—and therefore unintelligent—causes. Human agency in particular supervenes on underlying natural processes, which in turn usually are identified with brain function.

It is important to distinguish the naturalist's understanding of causation from the theist's. Within theism God is the ultimate reality. Consequently, whenever God acts, there can be nothing outside of God that compels God's action. God is not a billiard ball that must move when another billiard ball strikes it. God's actions are free, and though he responds to his creation, he does not do so out of necessity. Within theism, therefore, divine action is not reducible to some more basic mode of causation. Indeed, within theism divine action is the most basic mode of causation since any other mode of causation involves creatures which themselves were created in a divine act.

Now consider naturalism. Within naturalism nature is the ultimate reality. Consequently, whenever something happens in nature, there can be nothing outside of nature that shares responsibility for what happened. Thus, when an event happens in nature, it is either because some other event in nature was responsible for it or because it simply happened, apart from any other determining event. Events therefore happen either because they were caused by other events or because they happened spontaneously. The first of these is usually called "necessity," the second "chance." For the naturalist chance and necessity are the fundamental modes of causation. Together they constitute what are called "natural causes." Naturalism, therefore, seeks to account for intelligent agency in terms of natural causes.

How well have natural causes been able to account for intelligent agency? Cognitive scientists have achieved nothing like a full reduction. The French Enlightenment thinker Pierre Cabanis once remarked: *"Les nerfs—voilà tout l'homme"* (the nerves—that's all there is to man). A full reduction of intelligent agency to natural causes would give a complete account of human behavior, intention, and emotion in terms of neural processes. Nothing like this has been achieved. No doubt, neural processes are correlated with behavior, intention, and emotion. Anger presumably is correlated with certain localized brain excitations. But localized brain excitations hardly explain anger any better than do overt behaviors associated with anger—like shouting obscenities.

Because cognitive scientists have yet to effect a full reduction of intelligent agency to natural causes, they speak of intelligent agency as *supervening* on natural causes. Supervenience is a hierarchical relationship between higher order processes (in this case intelligent agency) and lower order processes (in this case natural causes). What supervenience says is that the relationship between the higher and lower order processes is a one-way street, with the lower determining the higher. To say, for instance, that intelligent agency supervenes on neurophysiology is to say that once all the facts about neurophysiology are in place, all the facts about intelligent agency are determined as well. Supervenience makes no pretense at reductive analysis. It simply asserts that the lower level determines the higher level—how it does it, we don't know.

Supervenience is therefore an insulating strategy, designed to protect a naturalistic account of intelligent agency until a full reductive explanation is found. Supervenience, though not providing a reduction, tells us that in principle a reduction exists. Given that nothing like a full reductive explanation of intelligent agency is at hand, why should we think that such a reduction is even possible? To be sure, if we knew that naturalism were correct, then supervenience would follow. But naturalism itself is at issue.

Neuroscience, for instance, is nowhere near achieving its ambitions, and that despite its strident rhetoric. Hardcore neuroscientists, for instance, refer disparagingly to the ordinary psychology of beliefs, desires, and emotions as "folk psychology." The implication is that just as "folk medicine" had to give way to "real medicine," so "folk psychology" will have to give way to a revamped psychology that is grounded in neuroscience. In place of taking cures that address our beliefs, desires, and emotions, tomorrow's healers of the soul will manipulate brain states directly and ignore such outdated categories as beliefs, desires, and emotions.

At least so the story goes. Actual neuroscience research has yet to keep pace with its vaulting ambition. That should hardly surprise us. The neurophysiology of our brains is incredibly plastic and has proven notoriously difficult to correlate with intentional states. For instance, Louis Pasteur, despite suffering a cerebral accident, continued to enjoy a flourishing scientific career. When his brain was examined after he died, it was discovered that half the brain had completely atrophied. How does one explain a flourishing intellectual life despite a severely damaged brain if mind and brain coincide?

Or consider a still more striking example. The December 12th, 1980 issue of Science contained an article by Roger Lewin titled "Is Your Brain Really Necessary?" In the article, Lewin reported a case study by John Lorber, a British neurologist and professor at Sheffield University. I quote from the article:

"There's a young student at this university," says Lorber, "who has an IQ of 126, has gained a first-class honors degree in mathematics, and is socially completely normal. And yet the boy has virtually no brain." [Lewin continues:] The student's physician at the university noticed that the youth had a slightly larger than normal head, and so referred him to Lorber, simply out of interest. "When we did a brain scan on him," Lorber recalls, "we saw that instead of the normal 4.5 centimeter thickness of brain tissue between the ventricles and the cortical surface, there was just a thin layer of mantle measuring a millimeter or so. His cranium is filled mainly with cerebrospinal fluid."

Against such anomalies, Cabanis's dictum, "the nerves—that's all there is to man," hardly inspires confidence. Yet as Thomas Kuhn has taught us, a science that is progressing fast and furiously is not about to be derailed by a few anomalies. Neuroscience is a case in point. For all the obstacles it faces in trying to reduce intelligent agency to natural causes, neuroscience persists in the Promethean determination to show that mind does ultimately reduce to neurophysiology. Absent a prior commitment to naturalism, this determination will seem misguided. On the other hand, given a prior commitment to naturalism, this determination is readily understandable.

Understandable yes, obligatory no. Most cognitive scientists do not rest their hopes with neuroscience. Yes, if naturalism is correct, then a reduction of intelligent agency to neurophysiology is in principle possible. The sheer difficulty of even attempting this reduction, both experimental and theoretical, however, leaves many cognitive scientists looking for a more manageable field to invest their energies in. As it turns out, the field of choice is computer science, and especially its sub-discipline of artificial intelligence (abbreviated AI). Unlike brains, computers are neat and precise. Also, unlike brains, computers and their programs can be copied and mass-produced. Inasmuch as science thrives on replicability and control, computer science offers tremendous practical advantages over neurological research.

Whereas the goal of neuroscience is to reduce intelligent agency to neurophysiology, the goal of artificial intelligence is to reduce intelligent agency to computer algorithms. Since computers operate deterministically, reducing intelligent agency to computer algorithms would indeed constitute a naturalistic reduction of intelligent agency. Should artificial intelligence succeed in reducing intelligent agency to computation, cognitive scientists would still have the task of showing in what sense brain function is computational (alternatively, Marvin Minsky's dictum "the mind is a computer made of meat" would still need to be verified). Even so, the reduction of intelligent agency to computation would go a long way toward establishing a purely naturalistic basis for human cognition.

An obvious question now arises: Can computation explain intelligent agency? First off, let's be clear that no actual computer system has come anywhere near to simulating the full range of capacities we associate with human intelligent agency. Yes, computers can do certain narrowly circumscribed tasks exceedingly well (like play chess). But require a computer to make a decision based on incomplete information and calling for common sense, and the computer will be lost. Perhaps the toughest problem facing artificial intelligence researchers is what's called the *frame problem*. The frame problem is getting a computer to find the appropriate frame of reference for solving a problem.

Consider, for instance, the following story: A man enters a bar. The bartender asks, "What can I do for you?" The man responds, "I'd like a glass of water." The bartender pulls out a gun and shouts, "Get out of here!" The man says "thank you" and leaves. End of story. What is the appropriate frame of reference? No, this isn't a story by Franz Kafka. The key item of information needed to make sense of this story is this: The man has the hiccups. By going to the bar to get a drink of water, the man hoped to cure his hiccups. The bartender, however, decided on a more radical cure. By terrifying the man with a gun, the bartender cured the man's hiccups immediately. Cured of his hiccups, the man was grateful and left. Humans are able to understand the appropriate frame of reference for such stories immediately. Computers, on the other hand, haven't a clue.

Ah, but just wait. Give an army of clever programmers enough time, funding, and computational power, and just see if they don't solve the frame problem. Naturalists are forever issuing such promissory notes, claiming that a conclusive confirmation of naturalism is right around the corner—just give our scientists a bit more time and money. John Polkinghorne refers to this practice as "promissory materialism."

Confronted with such promises, what's a theist to do? To refuse such promissory notes provokes the charge of obscurantism, but to accept them means suspending one's theism. It is possible to reject promissory materialism without meriting the charge of obscurantism. The point to realize is that a promissory note need only be taken seriously if there is good reason to think that it can be paid. The artificial intelligence community has thus far offered no compelling reason for thinking that it will ever solve the frame problem. Indeed, computers that employ common sense to determine appropriate frames of reference continue utterly to elude computer scientists.

Given the practical difficulties of producing a computer that faithfully models human cognition, the hardcore artificial intelligence advocate can change tactics and argue on theoretical grounds that humans are simply disguised computers. The argument runs something like this. Human beings are finite. Both the space of possible human behaviors and the space of possible sensory inputs are finite. For instance, there are only so many distinguishable word combinations that we can utter and only so many distinguishable sound combinations that can strike our eardrums. When represented mathematically, the total number of human lives that can be distinguished empirically is finite. Now it is an immediate consequence of recursion theory (the mathematical theory that undergirds computer science) that any operations and relations on finite sets are computable. It follows that human beings can be represented computationally. Humans are therefore functionally equivalent to computers. QED.

This argument can be nuanced. For instance, we can introduce a randomizing element into our computations to represent quantum indeterminacy. What's important here, however, is the gist of the argument. The argument asks us to grant that humans are essentially finite. Once that assumption is granted, recursion theory tells us that everything a finite being does is computable. We may never actually be able to build the machines that render us computable. But in principle we could, given enough memory and fast enough processors.

It's at this point that opponents of computational reductionism usually invoke Gödel's incompleteness theorem. Gödel's theorem is said to refute computational reductionism by showing that humans can do things that computers cannot—namely, produce a Gödel sentence. John Lucas made such an argument in the early 1960s, and his argument continues to be modified and revived. Now it is perfectly true that humans can produce Gödel sentences for computational systems external to themselves. But computers can as well be programmed to compute Gödel sentences for computational systems external to themselves. This point is seldom appreciated, but becomes evident from recursion-theoretic proofs of Gödel's theorem (see, for example, Klaus Weihrauch's *Computability*).

The problem, then, is not to find Gödel sentences for computational systems external to oneself. The problem is for an agent to examine oneself as a computational system and therewith produce one's own Gödel sentence. If human beings are non-computational, then there won't be any Gödel sentence to be found. If, on the other hand, human beings are computational, then, by Gödel's theorem, we won't be able to find our own Gödel sentences. And indeed, we haven't. Our inability to translate neurophysiology into computation guarantees that we can't even begin computing our Gödel sentences if indeed we are computational systems. Yes, for a computational system laid out before us we can determine its Gödel sentence. Nevertheless, we don't have sufficient access to ourselves to lay ourselves out before ourselves and thereby determine our Gödel sentences. It follows that neither Gödel's theorem nor our ability to prove Gödel's theorem shows that humans can do things that computers cannot.

Accordingly, Gödel's theorem fails to refute the argument for computational reductionism based on human finiteness. To recap that argument, humans are finite because the totality of their possible behavioral outputs and possible sensory inputs is finite. Moreover, all operations and relations on finite sets are by recursion theory computable. Hence, humans are computational systems. This is the argument. What are we to make of it? Despite the failure of Gödel's theorem to block its conclusion, is there a flaw in the argument?

Yes there is. The flaw consists in identifying human beings with their behavioral outputs and sensory inputs. Alternatively, the flaw consists in reducing our humanity to what can be observed and measured. We are more than what can be observed and measured. Once, however, we limit ourselves to what can be observed and measured, we are necessarily in the realm of the finite and therefore computable. We can only make so many observations. We can only take so many measurements. Moreover, our measurements never admit infinite gradations (indeed, there's always some magnitude below which quantities become empirically indistinguishable). Our empirical selves are therefore essentially finite. It follows that unless our actual selves transcend our empirical selves, our actual selves will be finite as well—and therefore computational.

Roger Penrose understands this problem. In *The Emperor's New Mind* and in his more recent *Shadows of the Mind*, he invokes quantum theory to underwrite a non-computational view of brain and mind. Penrose's strategy is the same that we saw for Gödel's theorem: Find something humans can do that computers can't. There are plenty of mathematical functions that are non-computable. Penrose therefore appeals to quantum processes in the brain whose mathematical characterization employs non-computable functions.

Does quantum theory offer a way out of computational reductionism? I would say no. Non-computable functions are an abstraction. To be non-computable, functions have to operate on infinite sets. The problem, however, is that we have no observational experience of infinite sets or of the non-computable functions defined on them. Yes, the mathematics of quantum theory employs non-computable functions. But when we start plugging in concrete numbers and doing calculations, we are back to finite sets and computable functions.

Granted, we may find it convenient to employ non-computable functions in characterizing some phenomenon. But when we need to say something definite about the phenomenon, we must supply concrete numbers, and suddenly we are back in the realm of the computable. Non-computability exists solely as a mathematical abstraction—a useful abstraction, but an abstraction nonetheless. Precisely because our behavioral outputs and sensory inputs are finite, there is no way to test non-computability against experience. All scientific data are finite, and any mathematical operations we perform on that data are computable. Non-computable functions are therefore always dispensable, however elegant they may appear mathematically.

There is, however, still a deeper problem with Penrose's program to eliminate computational reductionism. Suppose we could be convinced that there are processes in the brain that are non-computational. For Penrose they are quantum processes, but whatever form they take, as long as they are natural processes, we are still dealing with a naturalistic reduction of mind. Computational reductionism is but one type of naturalistic reductionism—certainly the most extreme, but by no means the only one. Penrose's program offers to replace computational processes with quantum processes. Quantum processes, however, are as fully naturalistic as computational processes. In offering to account for mind in terms of quantum theory, Penrose is therefore still wedded to a naturalistic reduction of mind and intelligent agency.

It's time to ask the obvious question: Why should anyone want to make this reduction? Certainly, if we have a prior commitment to naturalism, we will want to make it. But apart from that commitment, why attempt it? As we've seen, neurophysiology hasn't a clue about how to reduce intelligent agency to natural causes (hence its continued retreat to concepts like supervenience, emergence, and hierarchy—concepts which merely cloak ignorance). We've also seen that no actual computational systems show any sign of reducing intelligent agency to computation. The argument that we are computational systems because the totality of our possible behavioral outputs and possible sensory inputs is finite holds only if we presuppose that we are nothing more than the sum of those behavioral outputs and sensory inputs. So too, Penrose's argument that we are naturalistic systems because some well-established naturalistic theory (in this case quantum theory) characterizes our neurophysiology holds only if the theory does indeed accurately characterize our neurophysiology (itself a dubious claim given the frequency with which scientific theories are overturned) and so long as we presuppose that we are nothing more than a system characterized by some naturalistic theory.

Bottom line: The naturalistic reduction of intelligent agency is not the conclusion of an empirically-based evidential argument, but merely a straightforward consequence of presupposing naturalism in the first place. Indeed, the empirical evidence for a naturalistic reduction of intelligent agency is wholly lacking. For instance, nowhere does Penrose write down the Schrödinger equation for someone's brain, and then show how actual brain states agree with brain states predicted by the Schrödinger equation. Physicists have a hard enough time writing down the Schrödinger equation for systems of a few interacting particles. Imagine the difficulty of writing down the Schrödinger equation for the multi-billion neurons that constitute each of our brains. It ain't going to happen. Indeed, the only thing these naturalistic reductions of intelligent agency have until recently had in their favor is Occam's razor. And even this naturalistic mainstay is proving small comfort. Indeed, recent developments in the theory of intelligent design are showing that intelligent agency cannot be reduced to natural causes. Let us now turn to these developments.

The Resurgence of Design

In arguing against computational reductionism, both John Lucas and Roger Penrose attempted to find something humans can do that computers cannot. For Lucas, it was to construct a Gödel sentence. For Penrose, it was finding in neurophysiology a non-computational quantum process. Neither of these refutations succeeds against computational reductionism, much less against a general naturalistic reduction of intelligent agency. Nevertheless, the strategy underlying these attempted refutations is sound, namely, to find something intelligent agents can do that natural causes cannot. We don't have to look far. All of us attribute things to intelligent agents that we wouldn't dream of attributing to natural causes. For instance, natural causes can throw scrabble pieces on a board, but cannot arrange the pieces into meaningful sentences. To obtain a meaningful arrangement requires an intelligent agent.

This intuition, that natural causes are too stupid to do the things that intelligent agents are capable of, has underlain the design arguments of past centuries. Throughout the centuries theologians have argued that nature exhibits features which nature itself cannot explain, but which instead require an intelligence that transcends nature. From Church fathers like Minucius Felix and Basil the Great (third and fourth centuries) to medieval scholastics like Moses Maimonides and Thomas Aquinas (twelfth and thirteenth centuries) to reformed thinkers like Thomas Reid and Charles Hodge (eighteenth and nineteenth centuries), we find theologians making design arguments, arguing from the data of nature to an intelligence operating over and above nature.

Design arguments are old hat. Indeed, design arguments continue to be a staple of philosophy and religion courses. The most famous of the design arguments is William Paley's watchmaker argument. According to Paley, if we find a watch in a field, the watch's adaptation of means to ends (that is, the adaptation of its parts to telling time) ensures that it is the product of an intelligence, and not simply the output of undirected natural processes. So too, the marvelous adaptations of means to ends in organisms, whether at the level of whole organisms, or at the level of various subsystems (Paley focused especially on the mammalian eye), ensure that organisms are the product of an intelligence.

Though intuitively appealing, Paley's argument had until recently fallen into disuse. This is now changing. In the last five years design has witnessed an explosive resurgence. Scientists are beginning to realize that design can be rigorously formulated as a scientific theory. What has kept design outside the scientific mainstream these last hundred and forty years is the absence of a precise criterion for distinguishing intelligent agency from natural causes. For design to be scientifically tenable, scientists have to be sure they can reliably determine whether something is designed. Johannes Kepler, for instance, thought the craters on the moon were intelligently designed by moon dwellers. We now know that the craters were formed naturally. It's this fear of falsely attributing something to design only to have it overturned later that has prevented design from entering science proper. With a precise criterion for discriminating intelligently from unintelligently caused objects, scientists are now able to avoid Kepler's mistake.

Before examining this criterion, I want to offer a brief clarification about the word "design." I'm using "design" in three distinct senses. First, I use it to denote the scientific theory that distinguishes intelligent agency from natural causes, a theory that increasingly is being referred to as "design theory" or "intelligent design theory" (IDT). Second, I use "design" to denote what it is about intelligently produced objects that enables us to tell that they are intelligently produced and not simply the result of natural causes. When intelligent agents act, they leave behind a characteristic trademark or signature. The scholastics used to refer to the "vestiges of creation." The Latin *vestigium* means footprint. It was thought that God, though not physically present, left his footprints throughout creation. Hugh Ross has referred to the "fingerprint of God." It is "design" in this sense—as a trademark, signature, vestige, or fingerprint—that our criterion for discriminating intelligently from unintelligently caused objects is meant to identify. Lastly, I use "design" to denote intelligent agency itself. Thus, to say that something is designed is to say that an intelligent agent caused it.

Let us now turn to my advertised criterion for discriminating intelligently from unintelligently caused objects. Although a detailed treatment of this criterion is technical and appears in my book *The Design Inference*,[1] the basic idea is straightforward and easily illustrated. Consider how the radio astronomers in the movie *Contact* detected an extra-terrestrial intelligence. This movie, which came out last summer and was based on a novel by Carl Sagan, was an enjoyable piece of propaganda for the SETI research program—the Search for Extra-Terrestrial Intelligence. To make the movie interesting, the SETI researchers had to find an extra-terrestrial intelligence (the actual SETI program has yet to be so fortunate).

How, then, did the SETI researchers in *Contact* find an extra-terrestrial intelligence? To increase their chances of finding an extra-terrestrial intelligence, SETI researchers have to monitor millions of radio signals from outer space. Many natural objects in space produce radio waves. Looking for signs of design among all these naturally produced radio signals is like looking for a needle in a haystack. To sift through the haystack, SETI researchers run the signals they monitor through computers programmed with pattern-matchers. So long as a signal doesn't match one of the pre-set patterns, it will pass through the pattern-matching sieve. If, on the other hand, it does match one of those patterns, then, depending on the pattern matched, the SETI researchers may have cause for celebration.

The SETI researchers in *Contact* did find a signal worthy of celebration, namely the sequence of prime numbers from 2 to 101, represented as a series of beats and pauses (2 = beat-beat-pause; 3 = beat-beat-beat-pause; 5 = beat-beat-beat-beat-beat-pause; etc.). The

1 William Dembski, *The Design Inference: Eliminating Chance through Small Probabilities*, Cambridge University Press, 1998.

SETI researchers in *Contact* took this signal as decisive confirmation of an extra-terrestrial intelligence. What is it about this signal that warrants us inferring design? Whenever we infer design, we must establish two things—complexity and specification. Complexity ensures that the object in question is not so simple that it can readily be explained by natural causes. Specification ensures that this object exhibits the type of pattern that is the signature of intelligence.

To see why complexity is crucial for inferring design, consider what would have happened if the SETI researchers had simply witnessed a single prime number—say the number 2 represented by two beats followed by a pause. It is a sure bet that no SETI researcher, if confronted with this three-bit sequence (beat-beat-pause), is going to contact the science editor at the *New York Times*, hold a press conference, and announce that an extra-terrestrial intelligence has been discovered. No headline is going to read, "Aliens Master the Prime Number Two!"

The problem is that two beats followed by a pause is too short a sequence (that is, has too little complexity) to establish that an extra-terrestrial intelligence with knowledge of prime numbers produced it. A randomly beating radio source might by chance just happen to output the sequence beat-beat-pause. The sequence of 1126 beats and pauses required to represent the prime numbers from 2 to 101, however, is a different story. Here the sequence is sufficiently long (that is, has enough complexity) to confirm that an extra-terrestrial intelligence could have produced it.

Even so, complexity by itself isn't enough to eliminate natural causes and detect design. If I flip a coin 1000 times, I'll participate in a highly complex (or what amounts to the same thing, highly improbable) event. Indeed, the sequence I end up flipping will be one of 10300 possible sequences. This sequence, however, won't trigger a design inference. Though complex, it won't exhibit a pattern characteristic of intelligence. In contrast, consider the sequence of prime numbers from 2 to 101. Not only is this sequence complex, but it also constitutes a pattern characteristic of intelligence. The SETI researcher who in the movie *Contact* first noticed the sequence of prime numbers put it this way: "This isn't noise, this has structure."

What makes a pattern characteristic of intelligence and therefore suitable for detecting design? The basic intuition distinguishing patterns that alternately succeed or fail to detect design is easily motivated. Consider the case of an archer. Suppose an archer stands fifty meters from a large wall with bow and arrow in hand. The wall, let's say, is sufficiently large that the archer cannot help but hit it. Now suppose each time the archer shoots an arrow at the wall, the archer paints a target around the arrow so that the arrow sits squarely in the bull's-eye. What can be concluded from this scenario? Absolutely nothing about the archer's ability as an archer. Yes, a pattern is being matched; but it is a pattern fixed only after the arrow has been shot. The pattern is thus purely ad hoc.

But suppose instead the archer paints a fixed target on the wall and then shoots at it. Suppose the archer shoots a hundred arrows, and each time hits a perfect bull's-eye. What can be concluded from this second scenario? Confronted with this second scenario we are obligated to infer that here is a world-class archer, one whose shots cannot legitimately be referred to luck, but rather must be referred to the archer's skill and mastery. Skill and mastery are of course instances of design.

The type of pattern where the archer fixes a target first and then shoots at it is common to statistics, where it is known as setting a *rejection region* prior to an experiment. In statistics, if the outcome of an experiment falls within a rejection region, the chance hypothesis supposedly responsible for the outcome is rejected. Now a little reflection makes clear that a pattern need not be given prior to an event to eliminate chance and implicate design. Consider, for instance, a cryptographic text that encodes a message. Initially it looks like a random sequence of letters. Initially we lack any pattern for rejecting natural causes and inferring design. But as soon as someone gives us the cryptographic key for deciphering the text, we see the hidden message. The cryptographic key provides the pattern we need for detecting design. Moreover, unlike the patterns of statistics, it is given after the fact.

Patterns therefore divide into two types, those that in the presence of complexity warrant a design inference and those that despite the presence of complexity do not warrant a design inference. The first type of pattern I call a *specification*, the second a *fabrication*. Specifications are the non-ad hoc patterns that can legitimately be used to eliminate natural causes and detect design. In contrast, fabrications are the ad hoc patterns that cannot legitimately be used to detect design. The distinction between specifications and fabrications can be made with full statistical rigor.

Complexity and specification together yield a criterion for detecting design. I call it the complexity-specification criterion. According to this criterion, we reliably detect design in something whenever it is both complex and specified. To see why the complexity-specification criterion is exactly the right instrument for detecting design, we need to understand what it is about intelligent agents that makes them detectable in the first place. The principal characteristic of intelligent agency is choice. Whenever an intelligent agent acts, it chooses from a range of competing possibilities.

This is true not just of humans, but of animals as well as of extra-terrestrial intelligences. A rat navigating a maze must choose whether to go right or left at various points in the maze. When SETI researchers attempt to discover intelligence in the extra-terrestrial radio transmissions they are monitoring, they assume an extra-terrestrial intelligence could have chosen any number of possible radio transmissions, and then attempt to match the transmissions they observe with certain patterns as opposed to others. Whenever a human being utters meaningful speech, a choice is made from a range of possible sound-combinations that might have been uttered. Intelligent agency always entails discrimination, choosing certain things, ruling out others.

Given this characterization of intelligent agency, the crucial question is how to recognize it. Intelligent agents act by making a choice. How then do we recognize that an intelligent agent has made a choice? A bottle of ink spills accidentally onto a sheet of paper; someone takes a fountain pen and writes a message on a sheet of paper. In both instances ink is applied to paper. In both instances one among an almost infinite set of possibilities is realized. In both instances a contingency is actualized and others are ruled out. Yet in one instance we ascribe agency, in the other chance.

What is the relevant difference? Not only do we need to observe that a contingency was actualized, but we ourselves need also to be able to specify that contingency. The contingency must conform to an independently given pattern, and we must be able independently to formulate that pattern. A random inkblot is unspecifiable; a message written with ink on paper is specifiable. Ludwig Wittgenstein in *Culture and Value* made essentially the same point:

"We tend to take the speech of a Chinese for inarticulate gurgling. Someone who understands Chinese will recognize *language* in what he hears. Similarly I often cannot discern the *humanity* in man."

In hearing a Chinese utterance, someone who understands Chinese not only recognizes that one from a range of all possible utterances was actualized, but is also able to specify the utterance as coherent Chinese speech. Contrast this with someone who does not understand Chinese. In hearing a Chinese utterance, someone who does not understand Chinese also recognizes that one from a range of possible utterances was actualized, but this time, because lacking the ability to understand Chinese, is unable to specify the utterance as coherent speech.

To someone who does not understand Chinese, the utterance will appear gibberish. Gibberish—the utterance of nonsense syllables uninterpretable within any natural language—always actualizes one utterance from the range of possible utterances. Nevertheless, gibberish, by corresponding to nothing we can understand in any language, also cannot be specified. As a result, gibberish is never taken for intelligent communication, but always for what Wittgenstein calls "inarticulate gurgling."

This actualizing of one among several competing possibilities, ruling out the rest, and specifying the one that was actualized encapsulates how we recognize intelligent agency, or equivalently, how we detect design. Experimental psychologists who study animal learning and behavior have known this all along. To learn a task an animal must acquire the ability to actualize behaviors suitable for the task as well as the ability to rule out behaviors unsuitable for the task. Moreover, for a psychologist to recognize that an animal has learned a task, it is necessary not only to observe the animal making the appropriate discrimination, but also to specify this discrimination.

Thus to recognize whether a rat has successfully learned how to traverse a maze, a psychologist must first specify which sequence of right and left turns conducts the rat out of the maze. No doubt, a rat randomly wandering a maze also discriminates a sequence of right and left turns. But by randomly wandering the maze, the rat gives no indication that it can discriminate the appropriate sequence of right and left turns for exiting the maze. Consequently, the psychologist studying the rat will have no reason to think the rat has learned how to traverse the maze.

Only if the rat executes the sequence of right and left turns specified by the psychologist will the psychologist recognize that the rat has learned how to traverse the maze. Now it is precisely the learned behaviors we regard as intelligent in animals. Hence it is no surprise that the same scheme for recognizing animal learning recurs for recognizing intelligent agency generally, to wit: actualizing one among several competing possibilities, ruling out the others, and specifying the one chosen.

Note that complexity is implicit here as well. To see this, consider again a rat traversing a maze, but now take a very simple maze in which two right turns conduct the rat out of the maze. How will a psychologist studying the rat determine whether it has learned to exit the maze? Just putting the rat in the maze will not be enough. Because the maze is so simple, the rat could by chance just happen to take two right turns, and thereby exit the maze. The psychologist will therefore be uncertain whether the rat actually learned to exit this maze, or whether the rat just got lucky.

But contrast this now with a complicated maze in which a rat must take just the right sequence of left and right turns to exit the maze. Suppose the rat must take one hundred appropriate right and left turns, and that any mistake will prevent the rat from exiting the maze. A psychologist who sees the rat take no erroneous turns and in short order exit the maze will be convinced that the rat has indeed learned how to exit the maze, and that this was not dumb luck.

This general scheme for recognizing intelligent agency is but a thinly disguised form of the complexity-specification criterion. In general, to recognize intelligent agency we must observe a choice among competing possibilities, note which possibilities were not chosen, and then be able to specify the possibility that was chosen. What's more, the competing possibilities that were ruled out must be live possibilities, and sufficiently numerous so that specifying the possibility that was chosen cannot be attributed to chance. In terms of complexity, this is just another way of saying that the range of possibilities is complex.

All the elements in this general scheme for recognizing intelligent agency (that is, choosing, ruling out, and specifying) find their counterpart in the complexity-specification criterion. It follows that this criterion formalizes what we have been doing right along when we recognize intelligent agency. The complexity-specification criterion pinpoints what we need to be looking for when we detect design.

The implications of the complexity-specification criterion are profound, not just for science, but also for philosophy and theology. The power of this criterion resides in its generality. It would be one thing if the criterion only detected human agency. But as we've seen, it detects animal and extra-terrestrial agency as well. Nor is it limited to intelligent agents that belong to the physical world. The fine-tuning of the universe, about which cosmologists make such a to-do, is both complex and specified and readily yields design. So too, Michael Behe's irreducibly complex biochemical systems readily yield design. The complexity-specification criterion demonstrates that design pervades cosmology and biology. Moreover, it is a transcendent design, not reducible to the physical world. Indeed, no intelligent agent who is strictly physical could have presided over the origin of the universe or the origin of life.

Unlike design arguments of the past, the claim that transcendent design pervades the universe is no longer a strictly philosophical or theological claim. It is also a fully scientific claim. There exists a reliable criterion for detecting design—the complexity-specification criterion. This criterion detects design strictly from observational features of the world. Moreover, it belongs to probability and complexity theory, not to metaphysics and theology. And although it cannot achieve logical demonstration, it is capable of achieving statistical justification so compelling as to demand assent. When applied to the fine-tuning of the universe and the complex, information-rich structures of biology, it demonstrates a design external to the universe. In other words, the complexity-specification criterion demonstrates transcendent design.

This is not an argument from ignorance. Just as physicists reject perpetual motion machines because of what they know about the inherent constraints on energy and matter, so too design theorists reject any naturalistic reduction of specified complexity because of what they know about the inherent constraints on natural causes. Natural causes are too stupid to keep pace with intelligent causes. We've suspected this all along. Intelligent design theory provides a rigorous scientific demonstration of this longstanding intuition. Let me stress, the complexity-specification criterion is not a principle that comes to us demanding our unexamined acceptance—it is not an article of faith. Rather, it is the outcome of a careful and sustained argument about the precise interrelationships between necessity, chance, and design (for the details, please refer to my monograph *The Design Inference*).

Demonstrating transcendent design in the universe is a scientific inference, not a philosophical speculation. Once we understand the role of the complexity-specification criterion in warranting this inference, several things follow immediately: (1) Intelligent agency is logically prior to natural causation and cannot be reduced to it. (2) Intelligent agency is fully capable of making itself known against the backdrop of natural causes. (3) Any science that systematically ignores design is incomplete and defective. (4) Methodological naturalism, the view that science must confine itself solely to natural causes, far from assisting scientific inquiry actually stifles it. (5) The scientific picture of the world championed since the Enlightenment is not just wrong but massively wrong. Indeed, entire fields of inquiry, especially in the human sciences, will need to be rethought from the ground up in terms of intelligent design.

The Creation of the World

I want now to take stock and consider where we are in our study of the act of creation. In the phrase "act of creation," so far I have focused principally on the first part of that phrase—the "act" part, or what I've also been calling "intelligent agency." I have devoted much of my article till now to contrasting intelligent agency with natural causes. In particular, I have argued that no empirical evidence supports the reduction of intelligent agency to natural causes. I have also argued that no good philosophical arguments support that reduction. Indeed, those arguments that do are circular, presupposing the very naturalism they are supposed to underwrite. My strongest argument against the sufficiency of natural causes to account for intelligent agency, however, comes from the complexity-specification criterion. This empirically-based criterion reliably discriminates intelligent agency from natural causes. Moreover, when applied to cosmology and biology, it demonstrates not only the incompleteness of natural causes, but also the presence of transcendent design.

Now, within Christian theology there is one and only one way to make sense of transcendent design, and that is as a divine act of creation. I want therefore next to focus on divine creation, and specifically on the creation of the world. My aim is to use divine creation as a lens for understanding intelligent agency generally. God's act of creating the world is the prototype for all intelligent agency (creative or not). Indeed, all intelligent agency takes its cue from the creation of the world. How so? God's act of creating the world makes possible all of God's subsequent interactions with the world, as well as all subsequent actions by creatures within the world. God's act of creating the world is thus the prime instance of intelligent agency.

Let us therefore turn to the creation of the world as treated in Scripture. The first thing that strikes us is the mode of creation. God speaks and things happen. There is something singularly appropriate about this mode of creation. Any act of creation is the concretization of an intention by an intelligent agent. Now in our experience, the concretization of an intention can occur in any number of ways. Sculptors concretize their intentions by chipping away at stone; musicians by writing notes on lined sheets of paper; engineers by drawing up blueprints; etc. But in the final analysis, all concretizations of intentions can be subsumed under language. For instance, a precise enough set of instructions in a natural language will tell the sculptor how to form the statue, the musician how to record the notes, and the engineer how to draw up the blueprints. In this way language becomes the universal medium for concretizing intentions.

In treating language as the universal medium for concretizing intentions, we must be careful not to construe language in a narrowly linguistic sense (for example, as symbol strings manipulated by rules of grammar). The language that proceeds from God's mouth in the act of creation is not some linguistic convention. Rather, as John's Gospel informs us, it is the divine *Logos*, the Word that in Christ was made flesh, and through whom all things were created. This divine *Logos* subsists in himself and is under no compulsion to create. For the divine *Logos* to be active in creation, God must *speak* the divine *Logos*. This act of speaking always imposes a self-limitation on the divine *Logos*. There is a clear analogy here with human language. Just as every English utterance rules out those statements in the English language that were not uttered, so every divine spoken word rules out those possibilities in the divine *Logos* that were not spoken. Moreover, just as no human speaker of

English ever exhausts the English language, so God in creating through the divine spoken word never exhausts the divine *Logos*.

Because the divine spoken word always imposes a self-limitation on the divine *Logos*, the two notions need to be distinguished. We therefore distinguish *Logos* with a capital "L" (that is, the divine *Logos*) from *logos* with a small "l" (that is, the divine spoken word). Lacking a capitalization convention, the Greek New Testament employs *logos* in both senses. Thus in John's Gospel we read that "the *Logos* was made flesh and dwelt among us." Here the reference is to the divine *Logos* who incarnated himself in Jesus of Nazareth. On the other hand, in the First Epistle of Peter we read that we are born again "by the *logos* of God." Here the reference is to the divine spoken word that calls to salvation God's elect.

Because God is the God of truth, the divine spoken word always reflects the divine *Logos*. At the same time, because the divine spoken word always constitutes a self-limitation, it can never comprehend the divine *Logos*. Furthermore, because creation is a divine spoken word, it follows that creation can never comprehend the divine *Logos* either. This is why idolatry—worshipping the creation rather than the Creator—is so completely backwards, for it assigns ultimate value to something that is inherently incapable of achieving ultimate value. Creation, especially a fallen creation, can at best reflect God's glory. Idolatry, on the other hand, contends that creation fully comprehends God's glory. Idolatry turns the creation into the ultimate reality. We've seen this before. It's called naturalism. No doubt, contemporary scientific naturalism is a lot more sophisticated than pagan fertility cults, but the difference is superficial. Naturalism is idolatry by another name.

We need at all costs to resist naturalistic construals of *logos* (whether *logos* with a capital "L" or a small "l"). Because naturalism has become so embedded in our thinking, we tend to think of words and language as purely contextual, local, and historically contingent. On the assumption of naturalism, humans are the product of a blind evolutionary process that initially was devoid not only of humans but also of any living thing whatsoever. It follows that human language must derive from an evolutionary process that initially was devoid of language. Within naturalism, just as life emerges from non-life, so language emerges from the absence of language.

Now it's certainly true that human languages are changing, living entities—one has only to compare the King James version of the Bible with more recent translations into English to see how much our language has changed in the last 400 years. Words change their meanings over time. Grammar changes over time. Even logic and rhetoric change over time. What's more, human language is conventional. What a word means depends on convention and can be changed by convention. For instance, there is nothing intrinsic to the word "automobile" demanding that it denote a car. If we go with its Latin etymology, we might just as well have applied "auto-mobile" to human beings, who are after all "self-propelling." There is nothing sacred about the linguistic form that a word assumes. For instance, "gift" in English means a present, in German it means poison, and in French it means nothing at all. And of course, words only make sense within the context of broader units of discourse like whole narratives.

For Christian theism, however, language is never purely conventional. To be sure, the assignment of meaning to a linguistic entity is conventional. Meaning itself, however, transcends convention. As soon as we stipulate our language conventions, words assume meanings and are no longer free to mean anything an interpreter chooses. The deconstructionist claim that "texts are indeterminable and inevitably yield multiple, irreducibly diverse interpretations" and that "there can be no criteria for preferring one reading to another" is therefore false. This is not to preclude that texts can operate at multiple levels of meaning and interpretation. It is, however, to say that texts are anchored to their meaning and not free to float about indiscriminately.

Deconstruction's error traces directly to naturalism. Within naturalism, there is no transcendent realm of meaning to which our linguistic entities are capable of attaching. As a result, there is nothing to keep our linguistic usage in check save pragmatic considerations, which are always contextual, local, and historically contingent. The watchword for pragmatism is expedience, not truth. Once expedience dictates meaning, linguistic entities are capable of meaning anything. Not all naturalists are happy with this conclusion. Philosophers like John Searle and D. M. Armstrong try simultaneously to maintain an objective realm of meaning and a commitment to naturalism. They want desperately to find something more than pragmatic considerations to keep our linguistic usage in check. Insofar as they pull it off, however, they are tacitly appealing to a transcendent realm of meaning (take, for instance,

Armstrong's appeal to universals). As Alvin Plantinga has convincingly argued, objective truth and meaning have no legitimate place within a pure naturalism. Deconstruction, for all its faults, has this in its favor: it is consistent in its application of naturalism to the study of language.

By contrast, *logos* resists all naturalistic reductions. This becomes evident as soon as we understand what *logos* meant to the ancient Greeks. For the Greeks *logos* was never simply a linguistic entity. Today when we think "word," we often think a string of symbols written on a sheet of paper. This is not what the Greeks meant by *logos*. *Logos* was a far richer concept for the Greeks. Consider the following meanings of *logos* from Liddell and Scott's Greek-English Lexicon:

> the word by which the inward thought is expressed (speech)
> the inward thought or reason itself (reason)
> reflection, deliberation (choice)
> calculation, reckoning (mathematics)
> account, consideration, regard (inquiry, -ology)
> relation, proportion, analogy (harmony, balance)
> a reasonable ground, a condition (evidence, truth)

Logos is therefore an exceedingly rich notion encompassing the entire life of the mind.

The etymology of *logos* is revealing. *Logos* derives from the root *l-e-g*. This root appears in the Greek verb *lego*, which in the New Testament typically means "to speak." Yet the primitive meaning of *lego* is to lay; from thence it came to mean to pick up and gather; then to select and put together; and hence to select and put together words, and therefore to speak. As Marvin Vincent remarks in his New Testament word studies: "*logos* is a collecting or collection both of things in the mind, and of words by which they are expressed. It therefore signifies both the outward form by which the inward thought is expressed, and the inward thought itself, the Latin *oratio* and *ratio*: compare the Italian *ragionare*, 'to think' and 'to speak'."

The root *l-e-g* has several variants. We've already seen it as *l-o-g* in *logos*. But it also occurs as *l-e-c* in *intellect* and *l-i-g* in *intelligent*. This should give us pause. The word intelligent actually comes from the Latin rather than from the Greek. It derives from two Latin words, the preposition *inter*, meaning between, and the Latin (not Greek) verb *lego*, meaning to choose or select. The Latin *lego* stayed closer to its Indo-European root meaning than its Greek cognate, which came to refer explicitly to speech. According to its etymology, intelligence therefore consists in *choosing between*.

We've seen this connection between intelligence and choice before, namely, in the complexity-specification criterion. Specified complexity is precisely how we recognize that an intelligent agent has made a choice. It follows that the etymology of the word *intelligent* parallels the formal analysis of intelligent agency inherent in the complexity-specification criterion. The appropriateness of the phrase *intelligent design* now becomes apparent as well. Intelligent design is a scientific research program that seeks to understand intelligent agency by investigating specified complexity. But specified complexity is the characteristic trademark of choice. It follows that *intelligent design* is a thoroughly apt phrase, signifying that design is inferred precisely because an intelligent agent has done what only an intelligent agent can do, namely, make a choice.

If *intelligent design* is a thoroughly apt phrase, the same cannot be said for the phrase *natural selection*. The second word in this phrase, selection, is of course a synonym for choice. Indeed, the *l-e-c* in *selection* is a variant of the *l-e-g* that in the Latin *lego* means to choose or select, and that also appears as *l-i-g* in *intelligence*. Natural selection is therefore an oxymoron. It attributes the power to choose, which properly belongs only to intelligent agents, to natural causes, which inherently lack the power to choose. Richard Dawkins's concept of the *blind watchmaker* follows the same pattern, negating with *blind* what is affirmed in *watchmaker*. That's why Dawkins opens his book *The Blind Watchmaker* with the statement: "Biology is the study of complicated things that give the appearance of having been designed for a purpose." Natural selection and blind watchmakers don't yield actual design, but only the appearance of design.

Having considered the role of *logos* in creating the world, I want next to consider its role in rendering the world intelligible. To say that God through the divine *Logos* acts as an intelligent agent to create the world is only half the story. Yes, there is a deep and fundamental connection between God as divine *Logos* and God as intelligent agent—indeed, the very words *logos* and *intelligence* derive from the same Indo-European root. The world, however, is more than simply the product of an intelligent agent. In addition, the world is intelligible.

We see this in the very first entity that God creates—light. With the creation of light, the world becomes a place that is conceptualizable, and to which values can properly be assigned. To be sure, as God increasingly orders the world through the process of creation, the number of things that can be conceptualized increases, and the values assigned to things become refined. But even with light for now the only created entity, it is possible to conceptualize light, distinguish it from darkness, and assign a positive value to light, calling it good. The world is thus not merely a place where God's intentions are fulfilled, but also a place where God's intentions are intelligible. Moreover, that intelligibility is as much moral and aesthetic as it is scientific.

God, in speaking the divine *Logos*, not only creates the world but also renders it intelligible. This view of creation has far reaching consequences. For instance, the fact-value distinction dissolves opposite God's act of creation—indeed, what is and what ought to be unite in God's original intention at creation. Consider too Einstein's celebrated dictum about the comprehensibility of the world. Einstein claimed: "The most incomprehensible thing about the world is that it is comprehensible." This statement, so widely regarded as a profound insight, is actually a sad commentary on naturalism. Within naturalism the intelligibility of the world must always remain a mystery. Within theism, on the other hand, anything other than an intelligible world would constitute a mystery.

God speaks the divine *Logos* to create the world, and thereby renders the world intelligible. This fact is absolutely crucial to how we understand human language, and especially human language about God. Human language is a divine gift for helping us to understand the world, and by understanding the world to understand God himself. This is not to say that we ever comprehend God, as in achieving fixed, final, and exhaustive knowledge of God. But human language does enable us to express accurate claims about God and the world. It is vitally important for the Christian to understand this point. Human language is not an evolutionary refinement of grunts and stammers formerly uttered by some putative apelike ancestors. We are creatures made in the divine image. Human language is therefore a divine gift that mirrors the divine *Logos*.

Consider what this conception of language does to the charge that biblical language is hopelessly anthropomorphic. We continue to have conferences in the United States with titles like "Reimagining God." The idea behind such titles is that all our references to God are human constructions and can be changed as human needs require new constructions. Certain feminist theologians, for instance, object to referring to God as father. God as father, we are told, is an outdated patriarchal way of depicting God that, given contemporary concerns, needs to be changed. "Father," we are told is a metaphor co-opted from human experience and pressed into theological service. No. No. No. This view of theological language is hopeless and destroys the Christian faith.

The concept "father" is not an anthropomorphism, nor is referring to God as father metaphorical. All instances of fatherhood reflect the fatherhood of God. It's not that we are taking human fatherhood and idealizing it into a divine father image *à la* Ludwig Feuerbach or Sigmund Freud. Father is not an anthropomorphism at all. It's not that we are committing an anthropomorphism by referring to God as father. Rather, we are committing a "theomorphism" by referring to human beings as fathers. We are never using the word "father" as accurately as when we attribute it to God. As soon as we apply "father" to human beings, our language becomes analogical and derivative.

We see this readily in Scripture. Jesus enjoins us to call no one father except God. Certainly Jesus is not telling us never to refer to any human being as "father." All of us have human fathers, and they deserve that designation. Indeed, the Fifth Commandment tells us explicitly to honor our human fathers. But human fathers reflect a more profound reality, namely, the fatherhood of God. Or consider how Jesus responds to a rich, young ruler who addresses him as "good master." Jesus shoots back, "Why do you call me good? There is no one good except God." Goodness properly applies to God. It's not an anthropomorphism to call God good. The goodness we attribute to God is not an idealized human goodness. God defines goodness. When we speak of human goodness, it is only as subordinate to the divine goodness.

This view, that human language is a divine gift for understanding the world and therewith God, is powerfully liberating. No longer do we live in a Platonic world of shadows from which we must escape if we are to perceive the divine light. No longer do we live in a Kantian world of phenomena that bars access to noumena. No longer do we live in a naturalistic world devoid of transcendence. Rather, the world and everything in it becomes a sacrament, radiating God's glory. Moreover, our language is capable of celebrating that glory by speaking truly about what God has wrought in creation.

The view that creation proceeds through a divine spoken word has profound implications not just for the study of human language, but also for the study of human knowledge, or what philosophers call epistemology. For naturalism, epistemology's primary problem is unraveling Einstein's dictum: "The most incomprehensible thing about the world is that it is comprehensible." How is it that we can have any knowledge at all? Within naturalism there is no solution to this riddle. Theism, on the other hand, faces an entirely different problematic. For theism the problem is not how we can have knowledge, but why our knowledge is so prone to error and distortion. The Judeo-Christian tradition attributes the problem of error to the fall. At the heart of the fall is alienation. Beings are no longer properly in communion with other beings. We lie to ourselves. We lie to others. And others lie to us. Appearance and reality are out of sync. The problem of epistemology within the Judeo-Christian tradition isn't to establish that we have knowledge, but instead to root out the distortions that try to overthrow our knowledge.

On the view that creation proceeds through a divine spoken word, not only does naturalistic epistemology have to go by the board, but so does naturalistic ontology. Ontology asks what are the fundamental constituents of reality. According to naturalism (and I'm thinking here specifically of the scientific naturalism that currently dominates Western thought), the world is fundamentally an interacting system of mindless entities (be they particles, strings, fields, or whatever). Mind therefore becomes an emergent property of suitably arranged mindless entities. Naturalistic ontology is all backwards. If creation and everything in it proceeds through a divine spoken word, then the entities that are created don't suddenly fall silent at the moment of creation. Rather they continue to speak.

I look at a blade of grass and it speaks to me. In the light of the sun, it tells me that it is green. If I touch it, it tells me that it has a certain texture. It communicates something else to a chinch bug intent on devouring it. It communicates something else still to a particle physicist intent on reducing it to its particulate constituents. Which is not to say that the blade of grass does not communicate things about the particles that constitute it. But the blade of grass is more than any arrangement of particles and is capable of communicating more than is inherent in any such arrangement. Indeed, its reality derives not from its particulate constituents, but from its capacity to communicate with other entities in creation and ultimately with God himself.

The problem of being now receives a straightforward solution: To be is to be in communion, first with God and then with the rest of creation. It follows that the fundamental science, indeed the science that needs to ground all other sciences, is communication theory, and not, as is widely supposed an atomistic, reductionist, and mechanistic science of particles or other mindless entities, which then need to be built up to ever greater orders of complexity by equally mindless principles of association, known typically as natural laws. Communication theory's object of study is not particles, but the information that passes between entities. Information in turn is just another name for *logos*. This is an information-rich universe. The problem with mechanistic science is that it has no resources for recognizing and understanding information. Communication theory is only now coming into its own. A crucial development along the way has been the complexity-specification criterion. Indeed, specified complexity is precisely what's needed to recognize information.

Information—the information that God speaks to create the world, the information that continually proceeds from God in sustaining the world and acting in it, and the information that passes between God's creatures—this is the bridge that connects transcendence and immanence. All of this information is mediated through the divine *Logos*, who is before all things and by whom all things consist (Colossians 1:17). The crucial breakthrough of the intelligent design movement has been to show that this great theological truth—that God acts in the world by dispersing information—also has scientific content. All information, whether divinely inputted or transmitted between creatures, is in principle capable of being detected via the complexity-specification criterion. Examples abound:

The fine-tuning of the universe and irreducibly complex biochemical systems are instances of specified complexity, and signal information inputted into the universe by God at its creation.

Predictive prophecies in Scripture are instances of specified complexity, and signal information inputted by God as part of his sovereign activity within creation.

Language communication between humans is an instance of specified complexity, and signals information transmitted from one human to another.

The positivist science of this and the last century was incapable of coming to terms with information. The science of the new millennium will not be able to avoid it. Indeed, we already live in an information age.

Creativity, Divine and Human

In closing this article, I want to ask an obvious question: Why create? Why does God create? Why do we create? Although creation is always an intelligent act, it is much more than an intelligent act. The impulse behind creation is always to offer oneself as a gift. Creation is a gift. What's more, it is a gift of the most important thing we possess—ourselves. Indeed, creation is the means by which a creator—divine, human, or otherwise—gives oneself in self-revelation. Creation is not the neurotic, forced self-revelation offered on the psychoanalyst's couch. Nor is it the facile self-revelation of idle chatter. It is the self-revelation of labor and sacrifice. Creation always incurs a cost. Creation invests the creator's life in the thing created. When God creates humans, he breathes into them the breath of life—God's own life. At the end of the six days of creation God is tired—he has to rest. Creation is exhausting work. It is drawing oneself out of oneself and then imprinting oneself on the other.

Consider, for instance, the painter Vincent van Gogh. You can read all the biographies you want about him, but through it all van Gogh will still not have revealed himself to you. For van Gogh to reveal himself to you, you need to look at his paintings. As the Greek Orthodox theologian Christos Yannaras writes: "We know the person of van Gogh, what is unique, distinct and unrepeatable in his existence, only when we see his paintings. There we meet a reason (*logos*) which is his only and we separate him from every other painter. When we have seen enough pictures by van Gogh and then encounter one more, then we say right away: This is van Gogh. We distinguish immediately the otherness of his personal reason, the uniqueness of his creative expression."

The difference between the arts and the sciences now becomes clear. When I see a painting by van Gogh, I know immediately that it is his. But when I come across a mathematical theorem or scientific insight, I cannot decide who was responsible for it unless I am told. The world is God's creation, and scientists in understanding the world are simply retracing God's thoughts. Scientists are not creators but discoverers. True, they may formulate concepts that assist them in describing the world. But even such concepts do not bear the clear imprint of their formulators. Concepts like energy, inertia, and entropy give no clue about who formulated them. Hermann Weyl and John von Neumann were both equally qualified to formulate quantum mechanics in terms of Hilbert spaces. That von Neumann, and not Weyl, made the formulation is now an accident of history. There's nothing in the formulation that explicitly identifies von Neumann. Contrast this with a painting by van Gogh. It cannot be confused with a Monet.

The impulse to create and thereby give oneself in self-revelation need not be grand, but can be quite humble. A homemaker arranging a floral decoration engages in a creative act. The important thing about the act of creation is that it reveal the creator. The act of creation always bears the signature of the creator. It is a sad legacy of modern technology, and especially the production line, that most of the objects we buy no longer reveal their maker. Mass production is inimical to true creation. Yes, the objects we buy carry brand names, but in fact they are largely anonymous. We can tell very little about their maker. Compare this with God's creation of the world. Not one tree is identical with another. Not one face matches another. Indeed, a single hair on your head is unique— there was never one exactly like it, nor will there ever be another to match it.

The creation of the world by God is the most magnificent of all acts of creation. It, along with humanity's redemption through Jesus Christ, are the two key instances of God's self-revelation. The revelation of God in creation is typically called general revelation whereas the revelation of God in redemption is typically called special revelation. Consequently, theologians sometimes speak of two books, the Book of Nature, which is God's self-revelation in creation, and the Book of Scripture, which is God's self-revelation in redemption. If you want to know who God is, you need to know God through both creation and redemption. According to Scripture, the angels praise God chiefly for two things: God's creation of the world and God's redemption of the world through Jesus Christ. Let us follow the angels' example.

17

Epilogue

E. F. Schumacher

In the excitement over the unfolding of his scientific and technical powers, modern man has built a system of production that ravishes nature and a type of society that mutilates man. If only there were more and more wealth, everything else, it is thought, would fall into place. Money is considered to be all-powerful; if it could not actually buy non-material values, such as justice, harmony, beauty, or even health, it could circumvent the need for them or compensate for their loss. The development of production and the acquisition of wealth have thus become the highest goals of the modern world in relation to which all other goals, no matter how much lip-service may still be paid to them, have come to take second place. The highest goals require no justification; all secondary goals have finally to justify themselves in terms of the service their attainment renders to the attainment of the highest.

This is the philosophy of materialism, and it is this philosophy or metaphysic which is now being challenged by events. There has never been a time, in any society in any part of the world, without its sages and teachers to challenge materialism and plead for a different order of priorities. The languages have differed, the symbols have varied, yet the message has always been the same: "Seek ye *first* the kingdom of God, and all these things [the material things which you also need] shall be *added* unto you." They shall be added, we are told, here on earth where we need them, not simply in an after-life beyond our imagination. Today, however, this message reaches us not solely from the sages and saints but from the actual course of physical events. It speaks to us in the language of terrorism, genocide, breakdown, pollution, exhaustion. We live, it seems, in a unique period of convergence. It is becoming apparent that there is not only a promise but also a threat in those astonishing words about the kingdom of God—the threat that "unless you seek first the kingdom, these other things, which you also need, will cease to be available to you." As a recent writer put it, without reference to economics and politics but nonetheless with direct reference to the condition of the modern world:

> If it can be said that man collectively shrinks back more and more from the Truth, it can also be said that on all sides the Truth is closing in more and more upon man. It might almost be said that, in order to receive a touch of It, which in the past required a lifetime of effort, all that is asked of him now is not to shrink back. And yet how difficult that is![1]

We shrink back from the truth if we believe that the destructive forces of the modern world can be "brought under control" simply by mobilizing more resources—of wealth, education, and research—to fight pollution, to preserve wildlife, to discover new sources of energy, and to arrive at more effective agreements on peaceful coexistence. Needless to say, wealth, education, research, and many other things are needed for any civilization, but what is most needed today is a revision of the ends which these means are meant to serve. And this implies, above all else, the development of a lifestyle which accords to material things their proper, legitimate place, which is secondary and not primary.

The "logic of production" is neither the logic of life nor that of society. It is a small and subservient part of both. The destructive forces unleashed by it cannot be brought under control, unless the "logic of production" itself is brought under control—so that destructive forces cease to be unleashed. It is of little use trying to suppress terrorism if the production of deadly devices continues to be deemed a legitimate employment of man's creative powers. Nor can the fight against pollution be successful if the patterns of production and consumption continue to be of a scale, a complexity and a degree of violence which, as is becoming more and more apparent, do not fit into the laws of the universe, to which man is just as much subject as the rest of creation. Equally, the chance of mitigating the rate of resource depletion or of bringing harmony into the relationships between those in possession of wealth and

---

1 *Ancient Beliefs and Modern Superstitions* by Martin Lings (Perennial Books, London, 1964).

power and those without is non-existent as long as there is no idea anywhere of enough being good and more-than-enough being evil.

It is a hopeful sign that some awareness of these deeper issues is gradually—if exceedingly cautiously—finding expression even in some official and semi-official utterances. A report, written by a committee at the request of the Secretary of State for the Environment, talks about buying time during which technologically developed societies have an opportunity "to revise their values and to change their political objectives."[2] It is a matter of "moral choices," says the report; "no amount of calculation can alone provide the answers. . . . The fundamental questioning of conventional values by young people all over the world is a symptom of the widespread unease with which our industrial civilization is increasingly regarded."[3] Pollution must be brought under control and mankind's population and consumption of resources must be steered towards a permanent and sustainable equilibrium. "Unless this is done, sooner or later—and some believe that there is little time left—the downfall of civilization will not be a matter of science fiction. It will be the experience of our children and grandchildren."[4]

But how is it to be done? What are the "moral choices"? Is it just a matter, as the report also suggests, of deciding "how much we are willing to pay for clean surroundings?" Mankind has indeed a certain freedom of choice: it is not bound by trends, by the "logic of production," or by any other fragmentary logic. But it is bound by truth. Only in the service of truth is perfect freedom, and even those who today ask us "to free our imagination from bondage to the existing system"[5] fail to point the way to the recognition of truth.

It is hardly likely that twentieth-century man is called upon to discover truth that has never been discovered before. In the Christian tradition, as in all genuine traditions of mankind, the truth has been stated in religious terms, a language which has become well-nigh incomprehensible to the majority of modern

2 *Pollution: Nuisance or Nemesis?* (HMSO, London, 1972).
3 *Ibid.*
4 *Ibid.*
5 *Ibid.*

men. The language can be revised, and there are contemporary writers who have done so, while leaving the truth inviolate. Out of the whole Christian tradition, there is perhaps no body of teaching which is more relevant and appropriate to the modern predicament than the marvelously subtle and realistic doctrines of the Four Cardinal Virtues *prudentia, justitia, fortitudo,* and *temperantia.* The meaning of *prudentia,* significantly called the "mother" of all other virtues—*prudentia dicitur genitrix virtutum*—is not conveyed by the word "prudence," as currently used. It signifies the opposite of a small, mean, calculating attitude to life, which refuses to see and value anything that fails to promise an immediate utilitarian advantage.

> The pre-eminence of prudence means that realization of the good presupposes knowledge of reality. He alone can do good who knows what things are like and what their situation is. The pre-eminence of prudence means that so-called "good intentions" and so-called "meaning well" by no means suffice. Realization of the good presupposes that our actions are appropriate to the real situation, that is to the concrete realities which form the "environment" of a concrete human action; and that we therefore take this concrete reality seriously, with clear-eyed objectivity.[6]

This clear-eyed objectivity, however, cannot be achieved and prudence cannot be perfected except by an attitude of "silent contemplation" of reality, during which the egocentric interests of man are at least temporarily silenced.[7]

Only on the basis of this magnanimous kind of prudence can we achieve justice, fortitude, and *temperantia,* which means knowing when enough is enough. "Prudence implies a transformation of the knowledge of truth into decisions corresponding to reality." What, therefore, could be of greater importance today than the study and cultivation of prudence, which would almost inevitably lead to a real understanding of the three other cardinal virtues, all of which are indispensable for the survival of civilization?[8]

---

6 *Prudence* by Joseph Pieper, translated by Richard and Clara Winston (Faber & Faber Ltd., London, 1960).

7 *Fortitude and Temperance* by Joseph Pieper, translated by Daniel F. Coogan (Faber & Faber Ltd., London, 1955).

8 *Justice* by Joseph Pieper, translated by Lawrence E. Lynch (Faber & Faber Ltd., London, 1957). No better guide to the matchless Christian teaching of the Four

Justice relates to truth, fortitude to goodness, and *temperantia* to beauty; while prudence, in a sense, comprises all three. The type of realism which behaves as if the good, the true, and the beautiful were too vague and subjective to be adopted as the highest aims of social or individual life, or were the automatic spin-off of the successful pursuit of wealth and power, has been aptly called "crackpot realism." Everywhere people ask: "What can I actually do?" The answer is as simple as it is disconcerting: we can, each of us, work to put our own inner house in order. The guidance we need for this work cannot be found in science or technology, the value of which utterly depends on the ends they serve; but it can still be found in the traditional wisdom of mankind.

Cardinal Virtues could be found than Joseph Pieper, of whom it has been rightly said that he knows how to make what he has to say not only intelligible to the general reader but urgently relevant to the reader's problems and needs.

Acknowledgments

We would like to thank the following authors, editors and publishers for their consent to publish the articles in this anthology.

Frithjof Schuon, "**In the Wake of the Fall**"
Light on the Ancient Worlds (World Wisdom, Bloomington, 1984), pp. 28-57.

René Guénon, "**Sacred and Profane Science**"
The Crisis of the Modern World (Sophia Perennis et Universalis, Ghent, 2001), pp. 37-50.

Titus Burckhardt, "**Traditional Cosmology and the Modern World**"
Mirror of the Intellect (State University of New York Press, Albany, 1987), pp. 13-45.

Lord Northbourne, "**Religion and Science**"
Studies in Comparative Religion, Vol. 3, No. 4
(Perennial Books, Bedfont, Middlesex, 1969), pp. 142-154.

Seyyed Hossein Nasr, "**Contemporary Man, between the Rim and the Axis**"
Studies in Comparative Religion, Vol. 7, No. 2, (Perennial Books, Bedfont, Middlesex, 1973), pp. 113-126.

Philip Sherrard, "**Christianity and the Religious Thought of C. G. Jung**"
Christianity: Lineaments of a Sacred Tradition (Holy Cross Orthodox Press, Brookline, 1998), pp. 134-157.

James S. Cutsinger, "**On Earth as It Is in Heaven**"
Sacred Web: A Journal of Tradition and Modernity, Vol. 1 (July 1998), pp. 91-114.

Osman Bakar, "**The Nature and Extent of Criticism of Evolutionary Theory**"
Critique of Evolutionary Theory, Osman Bakar, ed.
(Kuala Lumpur: The Islamic Academy of Science and Nurin Enterprise), 1987, pp. 123-149.

Acknowledgments

D. M. Matheson, "**Knowledge and Knowledge**"
Tomorrow (Perennial Books, Bedfont, Middlesex, 1964),
pp. 115-118.

Gai Eaton, "**Knowledge and Its Counterfeits**"
King of the Castle (The Islamic Texts Society,
Cambridge, 1990), pp. 142-164.

Wendell Berry, "**Ignorance**"
Life is a Miracle: An Essay against Modern Superstition
(Counterpoint Press, New York, 2000), pp. 3-12.

Wolfgang Smith, "**The Plague of Scientistic Belief**"
Homiletic and Pastoral Review
(Ignatius Press, San Francisco, 2000).

Huston Smith, "**Scientism: The Bedrock of the Modern
Worldview**"
*Why Religion Matters: The Fate of the Human Spirit in an
Age of Disbelief* (Harper, San Francisco, 2000), pp. 59-78.

Giuseppe Sermonti, "**Life as Non-Historical Reality**"
Critique of Evolutionary Theory, Osman Bakar, ed.,
(The Islamic Academy of Science and Nurin Enterprise,
Kuala Lumpur, 1987), pp. 87-100.

Michael Robert Negus, "**Man, Creation and the Fossil Record**"
This is the first edition of the revised article originally
published in *Studies in Comparative Religion.*

William A. Dembski, "**The Act of Creation: Bridging
Transcendence and Immanence**"
(Strasbourg, presented as a lecture at Millstatt Forum,
1998).

E. F. Schumacher, "**Epilogue**"
Small Is Beautiful (Hartley & Marks Publishers, Inc.
by arrangement with HarperCollins, Vancouver, 1999),
pp. 248-252.

Throughout this anthology, the references to previously published
books have been updated to show the most recent editions in print.

Biographies of Contributors

Osman Bakar received his Bachelor of Science and Master of Science degrees in mathematics from the University of London and his Ph.D. in Islam from Temple University, Philadelphia. He is a former Professor of Philosophy of Science and Deputy Vice Chancellor (Academic) at the University of Malaya, Kuala Lumpur. He is one of the founding members and has also served as President of the Islamic Academy of Science of Malaysia. Dr. Bakar has written a dozen books and more than 100 articles on various aspects of Islamic thought and civilization, both classical and contemporary. He is currently Visiting Professor and Malaysia Chair of Islam in Southeast Asia at Georgetown University's School of Foreign Service in Washington, D.C.

Wendell Berry is a conservationist, farmer, essayist, novelist, professor of English and poet. He was born August 5, 1934 in Henry Country, Kentucky where he now lives on a farm. The *New York Times* has called Berry the "prophet of rural America." Mr. Berry is the author of 32 books of essays, poetry and novels. He has worked a farm in Henry County, Kentucky since 1965. He is a former professor of English at the University of Kentucky and a past fellow of both the Guggenheim Foundation and the Rockefeller Foundation. He has received numerous awards for his work, including an award from the National Institute and Academy of Arts and Letters in 1971, and most recently, the T. S. Eliot Award.

Titus Burckhardt (1908-1984) was a German Swiss and a contemporary of Frithjof Schuon, with whom he had a lifelong intellectual and spiritual friendship. Burckhardt's interests in philosophy, cosmology and the arts and crafts of traditional cultures came together to produce a unique literary corpus. In addition to his metaphysical works, he devoted a large portion of his writings to traditional cosmology, which he envisaged in a sense as the "handmaid of metaphysics." Some of his most notable works are: *An*

Introduction to Sufi Doctrine; Alchemy: Science of the Cosmos, Science of the Soul; Chartres and the Birth of the Cathedral; Fez: City of Islam. His works combine thorough scholarship with a synthetic intelligence that is particularly adept at relating principles to practice. Burckhardt's writings spanned a period of some 40 years, and they have continued to be published posthumously since his death.

James S. Cutsinger is currently Professor of Theology and Religious Thought at the University of South Carolina. He received his Ph.D. from Harvard University. A widely recognized authority on the *Sophia Perennis* and the traditionalist school of comparative religious thought, he is best known for his work on the philosopher Frithjof Schuon. The recipient of numerous teaching awards, he was honored in 1999 as a Michael J. Mungo University Teacher of the Year. Professor Cutsinger is a nationally known advocate of Socratic Teaching based on the classics. He is a frequent contributor to USC's Honors College, where he has taken the lead in developing a series of courses in the study of the Great Books. Among his works are: *Advice to the Serious Seeker: Meditations on the Teaching of Frithjof Schuon* and *Reclaiming the Great Tradition: Evangelicals, Catholics, and Orthodox in Dialogue.*

William A. Dembski, a mathematician and a philosopher, is associate research professor in the conceptual foundations of science at Baylor University and a senior fellow with Discovery Institute's Center for the Renewal of Science and Culture in Seattle. Dr. Dembski previously taught at Northwestern University, the University of Notre Dame, and the University of Dallas. He has done postdoctoral work in mathematics at M.I.T., in physics at the University of Chicago, and in computer science at Princeton University. A graduate of the University of Illinois at Chicago where he earned a B.A. in psychology, an M.S. in statistics, and a Ph.D. in philosophy, he also received a doctorate in mathematics from the University of Chicago in 1988 and a Master of Divinity degree from Princeton Theological Seminary in 1996. He has held National Science Foundation graduate and postdoctoral fellowships. Dr. Dembski has published articles in mathematics, philosophy, and theology journals and is the author/editor of seven books.

Charles le Gai Eaton was born in Lausanne, Switzerland of British parents in 1921. He was educated at Charterhouse and Kings College, Cambridge. After wartime service in the British Intelligence Corps, his professional life has included diplomatic service, teaching and journalism and has taken him to four continents.

On the one hand, his essays challenge many of the operating principles of modern—and thus secular—societies. They do so by contrasting prevalent ideas about human nature with the traditional view of man as *pontifex* or "God's viceroy on earth." Writing on religion, philosophy, politics and society, Eaton maintains that people are always free inwardly to shape their ultimate destinies.

He currently works as a consultant to the Islamic Cultural Center in London.

René Guénon (1886-1951) was born in Blois, France. Guénon was the forerunner-cum-originator of the traditionalist school of thought. His books have become classics in their field and continue to have a strong impact on intellectuals of both East and West.

The content of his work can be divided into four main subjects, the first two of which are metaphysical doctrine and traditional principles. Into a Europe that was severing all moorings to its traditional spiritual foundations and rapidly becoming secularized, René Guénon reintroduced the intellectual certitudes of metaphysics. He did so largely through his books *Man and His Becoming according to the Vedanta* and also *A General Introduction to the Study of Hindu Doctrines. Fundamental Symbols,* a posthumous collection of articles, brings together his many insightful essays on traditional symbolism, a third area of interest. Finally, his critique of the modern world is fully set forth in the two volumes *The Crisis of the Modern World* and *The Reign of Quantity.*

René Guénon passed away in the adopted homeland of his later years in Cairo, in 1951.

Donald McLeod Matheson (1896-1979) was educated at Oxford Univeristy. His interests in traditional spirituality led him to travel widely in India and the Middle East, where he made contact with a number of the outstanding religious figures of his time. His fluency in French resulted in his becoming a translator for the works of the perennialist authors Frithjof Schuon and Titus Burckhardt. In addition, he contributed articles of his own authorship for the British journal *Studies in Comparative Religion.*

Giovanni Monastra is the Scientific Coordinator of INRAN which is the National Research Institute for Food and Nutrition in Italy. Prior to this, he did research work in the laboratories of a pharmaceutical company, carrying out scientific studies mainly in the field of immunological drugs. He is the author of many articles and two books, one on the origin of life (*Le origini della vita,* 2000), with several criticisms of Darwinism, and the second on transgenic food, in which he presents arguments against genetic engineering (*Maschera e volto degli OGM,* 2002).

Seyyed Hossein Nasr is one of the world's leading experts on Islamic science and comparative religion. Born in 1933 in Iran, he received his B.S. in physics from M.I.T. and his Ph.D. in the history of science from Harvard, where he began his teaching career while pursuing his graduate studies. The author of over 50 books and 500 articles, Dr. Nasr is a well-known intellectual figure in both the West and the Islamic world. An eloquent speaker with an impressive academic record, his career as a teacher, lecturer and scholar spans more than four decades. The combination of his extensive erudition and his familiarity with many cultures has made him one of the most respected figures in comparative religion, philosophy and inter-faith dialogue, particularly the encounter between Islam and the West. Among his many books, some of the most important are: *Man and Nature; Religion and the Order of Nature; Knowledge and the Sacred.*

Dr. Nasr is currently University Professor of Islamic Studies at George Washington University in Washington, D.C.

Michael Robert Negus received Bachelor of Science degrees in chemistry, biology and zoology from the University of Reading in England. He holds a Masters in computer science from the University of Birmingham and a Ph.D. in zoological parasitology from Reading. He has taught zoology and biological sciences at the university level since 1965 and for 15 years was the Head of the Science Department at Newman College of Higher Education. He is the author of several articles on his biological area of research interest as well as the interface between science and religion. In 1996, he was a joint winner of a Templeton Award for courses dealing with science and theology.

Lord Northbourne (1898-1982) was educated at Oxford and was for many years Provost of Wye College—the agricultural college of London University. Early in his professional life, he began writing articles from a perspective that eventually has become known as "traditionalist" or "perennialist." A number of these essays appeared in the British journal, *Studies in Comparative Religion,* and were later published as books. His two most important books are *Looking Back on Progress* and *Religion in the Modern World,* which are presently being reprinted for the current generation of readers.

E. F. Schumacher (1911-1977) was born in Germany in 1911. A Rhodes Scholar at Oxford in the 1930's, he fled back to England before the Second World War to avoid living under Nazism. Although he was interned as an enemy alien during the War, his extraordinary abilities were recognized, and he was able to help the British government with its economic and financial mobilization.

After the War, Schumacher worked as an economic advisor in a variety of posts. In 1955 he traveled to Burma as an economic consultant. While there, he developed the principles of what he called "Buddhist economics," based on the belief that good work was essential for proper human development and that "production from local resources for local needs is the most rational way of economic life." He also gained insights that led him to become a pioneer of what is now called appropriate technology: earth- and user-friendly technology matched to the scale of community life. Schumacher subsequently became a featured writer in the British Journal *Resurgence.* His best-selling book, *Small Is Beautiful* (1973), was republished by Hartley & Marks in 1999. His two other books are *Good Work* and *A Guide for the Perplexed.*

Frithjof Schuon (1907-1998) was born of German parents in Basle, Switzerland. He is best known as the foremost spokesmen of the *religio perennis* and as a philosopher in the metaphysical current of the *Vedanta,* Meister Eckhart and Plato. Schuon's books offer the most comprehensive presentation of the *philosophia perennis,* the timeless metaphysical truth underlying the diverse religions. Besides its literal meaning, every religion has an esoteric dimension, which is essential primordial and universal. Because these truths are permanent and transcendent, the perspective has been called "perennialist."

Schuon's active career spanned over 50 years, during which he wrote more than 20 books on metaphysics, comparative religion, traditional and sacred art. Among his most important works are: *The Transcendent Unity of Religions; Form and Substance in the Religions; Logic and Transcendence* and *Survey of Metaphysics and Esoterism.*

Schuon's writings have been consistently featured and reviewed in a wide range of publications around the world. They are available in World Wisdom's "Library of Perennial Philosophy" series.

Frithjof Schuon passed away on May 5, 1998.

Giuseppe Sermonti has for several years been one of the leading biologists in Italy. His contribution to this anthology first appeared in *Rivista di Biologia* 73:4 (1980) pp. 551-569. The journal is a quarterly publication on biological sciences, published by the *Instituzione della Rivista di Biologia* of the University of Perugia, and Professor Sermonti was for many years its editor. Currently retired, Sermonti was formerly Professor of Embryology at the University of Perugia.

Philip Sherrard (1922-1995) was educated at Cambridge. Among the works for which he is best known is his collaboration in the complete translation of the *Philokalia*. The combination of his interests in metaphysics, theology, art and aesthetics led to his participation in the review *Temenos*, of which he was one of the founders in 1980. In all his works, Dr. Sherrard sought to express an all-embracing vision, in which the natural and the supernatural come together in a wholeness that bears witness to the numinous wonder of life. This vision is the leit-motif of his last three books: *The Eclipse of Man and Nature; The Sacred in Life and Art;* and *Human Image, World Image.*

Philip Sherrard passed away on May 30, 1995.

Huston Smith is one of the most well-known figures in the field of comparative religion. He has had a long and active teaching career that has included posts at Washington University, M.I.T. and the University of California at Berkeley. He is currently the Thomas

J. Watson Professor of Religion and Distinguished Adjunct Professor of Philosophy, Emeritus, at Syracuse University.

Dr. Smith is the author of numerous articles that have appeared in professional and popular journals. His book *The World's Religions* (formerly *The Religions of Man*) has sold several million copies and has been the most widely used textbook for courses on comparative religion for many years. His most recent book is *Why Religion Matters: the Fate of the Human Spirit in an Age of Disbelief.* In it, he gives a well-grounded critique of modernism and argues convincingly for the restoration of religious belief as the primary stabilizing basis for individuals and for society.

Wolfgang Smith graduated at age 18 from Cornell University with a B.A. in mathematics, physics, and philosophy. Two years later he took an M.S. in theoretical physics at Purdue University. After receiving a Ph.D. in mathematics from Columbia University, Dr. Smith held professorial positions at M.I.T., U.C.L.A., and Oregon State University until his retirement in 1992. He has published extensively on mathematical topics relating to algebraic and differential topology.

From the start, however, Smith has evinced a dominant interest in metaphysics and theology. Early in life he acquired a taste for Plato and the Neoplatonists, and sojourned in India to gain acquaintance with the Vedantic tradition. Later he devoted himself to the study of theology, and began his career as a Catholic metaphysical author. Besides contributing numerous articles to scholarly journals, Dr. Smith has authored three books: *Cosmos and Transcendence* (1984), *Teilhardism and the New Religion* (1988), and *The Quantum Enigma* (1995).

About the Editor

Mehrdad Zarandi was born in 1963 in Kerman, Iran. From an early age, he had a keen interest in science and mathematics. His undergraduate liberal arts studies brought him into contact with the works of contemporary perennialist authors, whose perspective has had a profound effect on his intellectual formation. He earned Master of Science and Doctoral degrees in chemical engineering from the California Institute of Technology, where he continued to work as a research scientist. During this time he also held the position of adjunct professor of physics at Occidental College. He is currently a visiting faculty member in aeronautics and biomedical engineering at Caltech. In addition to his technical research and publications, he has enjoyed a wide range of teaching experiences with undergraduates in mathematics, chemistry, physics and the history and philosophy of science. His interest in the correspondences between metaphysical principles and their expression within cosmology and science led to his collecting the essays for this anthology.

Index